Geographies of Transport and Mobility

Geographies of Transport and Mobility aims to provide a comprehensive and evidenced account of the intellectual and pragmatic challenges for personal mobility in the twenty-first century. In doing so, it argues that geographers have a key role to play in shaping academic and policy debates on how personal mobility can become more sustainable.

The book is structured in three parts. Part I explores how personal mobility has evolved since the mid-nineteenth century, plotting the intricate relationship between new forms of mobile technology, urban planning and design and social practices. Part II examines how researchers study transport and mobility, and outlines the different intellectual trajectories of transport geography and geographies of mobilities. Part III then outlines and discusses the discourse of sustainable mobility that has emerged in recent years; the ways in which social, economic and environmental sustainability can be promoted through different strategies, focusing on behavioural change and urban design.

Geographies of Transport and Mobility provides a unique perspective on personal mobility by demonstrating how the way we travel has developed through complex economic and social processes. It argues that this historical context is critical for considering how mobility in the twenty-first century can be more sustainable, not just environmentally, but also economically and socially. As such, it argues for a renewed focus on sustainable place-making as a way to radically shift mobility practices. *Geographies of Transport and Mobility* is designed to appeal to advanced level undergraduate students and researchers in the fields of geography, anthropology, psychology, sociology and transport studies.

Stewart Barr is Professor of Geography and has worked as a researcher and lecturer at the University of Exeter since 2001. His current research focuses on critically understanding intellectual and policy discourses on behavioural change and sustainability.

Jan Prillwitz is an independent travel behaviour researcher who holds a PhD in Geography from Leipzig University. His main research interests are in sustainable travel, mobility styles, concepts of new mobilities and the role of socio-psychological factors for individual travel decisions.

Tim Ryley is Professor of Aviation at Griffith University, Brisbane, Australia, where he is head of the School of Natural Sciences. He has over 20 years' experience within transportation research and his expertise is in the fields of airport planning, airport surface access, airport operations and air travel demand, as well as the broader relationship between transport and climate change.

Gareth Shaw is Professor of Retail and Tourism Management at the University of Exeter Business School and is also currently an Innovation Fellow at the Advanced Institute of Management. He was formerly Professor of Human Geography at the University of Exeter and undertakes research on tourism innovation and tourist behaviour.

Transport and Mobility
Series Editor: John Nelson

The inception of this series marks a major resurgence of geographical research into transport and mobility. Reflecting the dynamic relationships between socio-spatial behaviour and change, it acts as a forum for cutting-edge research into transport and mobility, and for innovative and decisive debates on the formulation and repercussions of transport policy making.

For a full list of titles in this series, please visit www.routledge.com/Transport-and-Mobility/book-series/ASHSER-1188

Geographies of Transport and Mobility

Prospects and Challenges in an Age of Climate Change

Stewart Barr, Jan Prillwitz, Tim Ryley and Gareth Shaw

Routledge
Taylor & Francis Group

LONDON AND NEW YORK

First published 2018 by Routledge

2 Park Square, Milton Park, Abingdon, Oxfordshire OX14 4RN

52 Vanderbilt Avenue, New York, NY 10017

Routledge is an imprint of the Taylor & Francis Group, an informa business

First issued in paperback 2020

British Library Cataloguing in Publication Data
A catalogue record for this book is available from the British Library

Library of Congress Cataloging in Publication Data
Names: Barr, Stewart, 1976- author. | Prillwitz, Jan, co-author. | Ryley,
Tim, co-author. | Shaw, Gareth, co-author.
Title: Geographies of transport and mobility : prospects and challenges
in an age of climate change / Stewart Barr, Jan Prillwitz, Tim Ryley, and
Gareth Shaw.
Description: Milton Park, Abingdon, Oxon ; New York, NY : Routledge,
2018. | "Simultaneously published in the USA and Canada"--Title page
verso. | Includes bibliographical references and index.
Identifiers: LCCN 2017025613| ISBN 9781409447030 (hardback) |
ISBN 9781315584461 (ebook)
Subjects: LCSH: Transportation--Social aspects. | Transportation--
Environmental aspects. | Travel--Social aspects. | Travel--Environmental
aspects. | Climatic changes--Social aspects. | Human geography. |
Sustainability.
Classification: LCC HE151 .B367 2018 | DDC 388--dc23
LC record available at https://lccn.loc.gov/2017025613

ISBN: 978-1-4094-4703-0 (hbk)
ISBN: 978-0-367-36232-4 (pbk)

Typeset in Times New Roman
by Sunrise Setting Ltd, Brixham, UK

Contents

Figures

Tables

Preface

During the course of writing this book, the view from my office window has changed forever. What was once a large sloping field of green grass became initially an overflow car park and now a seemingly permanent monument to the overwhelming power of auto-mobility in our age. What is perhaps most striking is the lack of interest this change provoked. In contrast to the construction of new buildings (hotly contested with differing opinions on everything from aesthetics, visibility and the conservation of wildlife), a new car park can often be regarded as a 'necessary evil' to keep the wheels of economic production turning. As such, building more roads, more lanes on the increasing number of roads, and more car parks on which to place the increasing number of cars generated has become closely linked to promoting economic growth. And if that growth results in more car journeys, then we should of course be building more roads, lanes, car parks and so on.

In this book my co-authors and I attempt to step back from this situation and to ask why auto-mobility has become such an all-pervasive 'solution' to our apparent mobility needs. In so doing, we argue that the car's prominent position in most Western societies has as much to do with how we use and plan space, as it does with either technology or our own individual decision-making. Accordingly, we seek to provide a geographical perspective on what has sometimes been treated as either a lack of 'smart' solutions or an inability to get people to do the 'right' thing in their travel choices. We show that the way we travel has much to do with the way we manage space as a society and how spaces of mobility have emerged over the last century and a half, as a result of how we plan, maintain and use the places we live, work and take leisure time.

In particular, we attempt to show how contemporary mobility is underpinned by two major forces. First, the spatial configuration of our cities, economic activities and leisure practices promote mobility as a necessity for living. Gone are the days when vast numbers of people lived near the factory in which they worked and had location-specific friendship networks. Today's highly distributed societies necessitate movement for economic transaction, employment, education and entertainment. Second, to be mobile is not only a necessity, but a desirable state of being and one that bestows status and can be used as cultural capital in an experience economy. Little wonder then that mobility, and the status it affords, is one of the signifiers of our age.

Why does personal mobility matter? There are perhaps two concerns here, one of which might seem obvious, but the second is often over-looked. First, reliance on the motor vehicle and commercial air travel has clear implications for anthropogenic climate change through carbon emissions. In essence, the projected growth in both personal car transport and flying over the coming decades poses a major challenge for maintaining annual mean temperatures at what are regarded safe levels. Indeed, there are clearly environmental concerns at the local scale; increased air transport and private vehicle usage causes reductions in air and noise quality and can lead to significant increases in highly localised health problems (such as respiratory conditions).

However, there is a second concern that we as geographers wish to highlight. This is the challenge posed by the apparent dis-connection from place that many now experience and which has been promoted through the need and desire for travel. We can see this in so many ways: frequent travel by car to buy bread and a pint of milk; the negotiation of traffic congestion as so many struggle to commute to work; the drive to an out-of-town shopping centre to buy anything from pet food to a new TV; the desire for ever more exotic experiences far away from home. All of this leads, we argue, to a dwindling concern for the places where we all live. We argue that paradoxically, the emergence of the private motor vehicle as the travel mode of choice has been one of the main reasons that the want and need to travel has increased. Years of post-war planning transformed some settlements in the UK and North America into dead zones of freeways, dystopian car parks and cut-off communities. Who would want to live in some of these places? Perhaps a better life could be found in suburbia. But of course, that necessitates private transport and all of the infrastructure and paraphernalia that accompanies this technology: freeways, inter-changes, petrol stations, service stations, shopping centres, drive-thru eating establishments, under-passes, flyovers, bridges and the associated impacts of noise, air pollution and congestion.

Yet examples highlighted in this book demonstrate that different models of building and sustaining place can and do exist. Whether it's Freiburg in Germany or Portland, Oregon, or much smaller but no less important examples, there are ways of planning places that make them truly liveable: streets for people not cars; buildings that are appropriately scaled and aesthetically diverse; abundant, affordable and integrated public transport; mixed-use developments that combines residential and commercial property; and diverse, vibrant street culture. The examples we showcase in this book demonstrate how such patterns of development can promote a transition to a public transit orientated society and one where quality of life is ranked consistently high.

So why doesn't this happen everywhere? The answer to this question is of course complex and has much to do with a deeply embedded culture of privileging private transport above public modes of travel, but it also has to do with an increasing reliance on technology, which can act to promote resistance to fundamentally changing behaviour and re-shaping places. As such, there is an obsession with finding more ways to maintain levels of private transport and mobility, without stopping to consider whether this is an appropriate strategy. In an emerging era of so-called 'smart cities' (Albino, 2015; Kitchin, 2015a), we might ask what the smarter choice is. Undoubtedly there are huge benefits to be gained from deploying and carefully using technology to help us achieve our goals – but what are these goals? We argue here that we should start by asking what kinds of places are desirable to live in. It's likely that our answer would have some of the characteristics we described above. And to achieve that kind of vision means that places need to be planned for that don't rely on private motor transport for their functioning. Such a change is ambitious and requires political leadership. It requires up-ending at least 70 years of planning doctrine driven by 'predict and provide'.

And what would the result look like? Perhaps it would be one where the number of trips reduced; where journeys were based on cycling, walking and mass transit; where the journey to buy a pint of milk was not a ride in the utility vehicle, and where commuting to work was not such a hassle. So, we ask our readers to look at where they live and work and ask: what would I like this place to be like and what would make me value it some more? We suspect that much of the answer would in fact be connected with how that place does or doesn't 'work'. And if it doesn't work, it's likely that the architectures and practices of mobility have much to do with it, as explored within the chapters of this book.

Acknowledgements

The authors would like to thank all those who have contributed to the production of this book. In particular, thanks are due to Lisa Cole of the University of Exeter's Design Studio, who prepared the maps and images. Dominic Walker and Matthew Cole helped to edit the text and assisted in formatting. Particular thanks are due to research participants, collaborators, colleagues and students who have provided encouragement during the book's production and who have helped to shape the intellectual arguments. The authors also acknowledge the funding received from the Economic and Social Research Council and Innovate UK via the Natural Environment Research Council, which has funded much of the research that forms the basis for the book (Grant numbers: ES/F00169X/1; ES/J001007/1; TS/N001672/1).

Part I

Contextualising geographies of transport and mobility

1   Geographies of transport

Geographies of mobility

A world of movement

It is hard for most of us to grasp the fact that had we lived just 200 years ago, most of us would have spent our lives living within a limited geographical area, often confined to the hamlet, village or immediate rural area in which we and our extended family lived. Travel, such as it was, was difficult, time consuming and restricted to those with the wealth to afford a stagecoach. Indeed, the modern notion of a road was yet to appear; most roads were dirt tracks with uneven surfaces, turning into boggy mires during winter.

How is it then, that in just 200 years or less, our very identities have become so profoundly framed by our ability to move and to do so at speed, with ease and in relative comfort? This book attempts to address this question and to illustrate how the way we move is deeply implicated in underlying social and political processes, which have fundamentally changed our relationship with space and our relationship to place – two key geographical concepts. In so doing, the book aims to consider how we have become locked into travel not only as a way that we do business, arrange our living environments and take leisure time, but also as a key manifestation of our post-modern identity. We are, in this sense, consumers of space. We crave new experiences through space and as such have entered a period of hyper-mobility, as the late John Urry (2011) put it, in which we are always mobile. Taking this intellectual position, we aim to demonstrate how this construction of ourselves as mobile citizens is having a range of social, economic and ecological impacts, and how future (im)mobilities may emerge from changes to the very resource base on which travel depends – fossil fuels, particularly oil.

In this introductory chapter, we begin by providing three important contexts for this book. First, we examine the meaning of transport and mobility as both popular and academic terms. In so doing, we will argue that the differences in these meanings provides a framework for understanding both the ways in which academics have approached the study of transport and mobility and the out-turn of policies intended to shape how we travel. Second, we situate the book within the key relationships that we see as critical for an understanding of contemporary mobility; that is between what is manifested as individual travel behaviour, the social construction of mobility and its manifestation as social practice, and the surrounding infrastructures and technologies in which we live and work. Finally, we argue that transport and mobility need to be viewed as dynamic and developing concepts, which are implicated in broader socio-political and economic processes and which are likely to change to reflect these processes. Accordingly, we contend that an understanding of transport and mobility for the future must be realistic, recognising that shifting the social practice of hyper-mobility is unlikely to be an easy or politically manageable task for the near future.

Transport and mobility

Thought of solely in a popular context, transport is often a term associated with specific technologies, such as the motor car, bus or train. Mobility, by contrast, is often regarded as something personal, an expression of one's ability to move both in real time and through life stages. Intellectually, this distinction is not entirely different and it is one that has shaped the landscape of transport geography for some time now. As we will examine in Chapter 4, the academic landscape of transport geography is one that straddles several epistemological traditions and these are reflected in the research and methodologies applied to study how and why people move in the way that they do.

Transport geography has a long and extensive disciplinary history that has its origins in the spatial turn of the 1960s and 1970s, and to some extent the legacy of this period of expansion has lasted to the present day. As Rodrigue *et al.* (2013) note, the geography of transport has largely focused on the analysis of networks and flows and has sought to examine how the movement of people and products through space can be described, explained and modelled. As such, transport geography has traditionally been focused on spatial analysis and an up-scaled interrogation of mobility. That is, it has been concerned with how transport systems function and the ways in which policy problems can be examined through a spatial focus.

This approach has necessarily led transport geographers to adopt a number of methodological approaches for their work, which have largely been focused on the use of various types of quantitative research tools and modelling approaches (Knowles *et al.*, 2008). Accordingly, the questions that transport geographers have sought to address are those which aim to explore the trends and patterns that characterise systems in space. Yet in recent years, social scientists have also become increasingly concerned with transport as a social entity. In this way, transport has been analysed not only as a technical or physical entity, but also from a cultural and social perspective. Known as the mobilities turn in the social sciences (see Urry, 2011), this research agenda has sought to explore the meanings associated with travel and the ways in which mobility is reflective of broader social and economic trends. In this way, social scientists have argued for situating transport studies within a broader framework that appreciates how mobility is a socially constructed phenomenon. As such, mobilities research is as interested in *why* people move as it is in *how* they move.

This approach has reconnected transport studies with the broader social sciences and has encouraged those, for example, exploring travel behaviour to consider how such individualised 'choices' are related to the underlying economic and social infrastructures that drive embedded and embodied practices of movement (Adey, 2010; Larsen *et al.*, 2012). Accordingly, this approach has adopted much more intensive and qualitative research methods, as scholars seek to uncover the complex social processes that (re)produce forms of mobility.

In this book, we aim to demonstrate how a socially embedded understanding of how and why we travel can have major benefits for appreciating our current travel predicament, and how we need to view the challenge of promoting sustainable mobility as one of changing underlying values, for which our current patterns of mobility are largely representative, rather than being discrete choices, isolated from other parts of our everyday life. In this way, we present here an alternative manifesto for analysis and change to that of the increasingly dominant 'smart cities' agenda, which is garnering attention amongst academic researchers, planners and technologically developers (Glasmeier and Christopherson, 2015).

The smart cities agenda takes as its starting point the need to make cities more efficient and to deploy macro-scale infrastructure-technology ('big data') and the use of 'everyware' to reduce congestion and promote 'better' decision-making amongst travellers (Kitchin, 2013, 2014, 2015c). Moreover, through integrating 'big data' with the 'everyware' of readily available mobile technologies, a kind of utopian vision for rational decision-making is being pursued, which is a recurrent theme in countries like the UK (Jones *et al.*, 2013). However, while acknowledging that there are likely to be benefits of carefully exploring the advantages of integrating technological developments alongside urban development and citizen engagement (Kitchin, 2014), we argue here that current manifestations of the smart city side-step how we can reshape cities to radically change mobility practices and to improve urban living. This is because many smart city programmes are governed by small elites that present a particular view of how cities ought to be managed, raising a range of ethical concerns (Hollands, 2015; Söderström *et al.*, 2014; Vanolo, 2014). As such, citizens are often positioned as passive consumers of technology, rather than being co-creators of new practices and spaces (Gabrys, 2014; Viitanen and Kingston, 2014) 'the smart city discourse may be a powerful tool for the production of docile subjects and mechanisms of political legitimisation' (Vanolo, 2014, p. 883). As such, current smart city agendas do little to challenge the conventional approach towards transport and city planning that has promoted mobility (Banister, 2008). In this way, smart city developments are frequently models *out of place* (Kitchin, 2015a), such that they don't start with a given city, but rather a model that is inflicted upon a space (Hollands, 2015). In this book we argue the opposite; we need to start with place and to appreciate the links between how spaces are configured and the ways that people move. Indeed, in making changes, we advocate approaches that are place-led and inclusive. Such changes will often involve technology, but we argue here that technology ought to serve the needs of people and places, rather than cities becoming the sites of experiments for technological utopians (Bulkeley and Castán Broto, 2013).

Situating transport and mobility

We take as our organising framework three critical constructs that form the basis of how we argue that transport and mobility needs to be understood. This framework has emerged from recent scholarship in the social sciences that has questioned the rationalistic ways in which social scientists have often viewed human decision-making and consequent acknowledgement of the role that social trends and broader political and economic framings have in understanding how we travel.

The three constructs we wish to explore derive from the landscape of transport geography and mobilities research. First, we acknowledge the important role of the individual as a frame of reference and unit of study within transport geography and positivistic social science scholarship more generally. There is a long tradition within transport geography of studying individual travel choices and doing so through the theoretical lens of psychology (Anable, 2005). To this end, the individual has been the lens through which travel behaviour has been understood; individuals, so the argument goes, make reasoned and rational choices, such as travel mode, based on a range of discrete factors. As Chapter 5 will highlight, this framing of individual travel choices has been reflected in the application of a range of conceptual tools to quantitatively explain travel-mode choices, such as Ajzen's (1991) Theory of Planned Behaviour. Through this intellectual lens, the way in which we travel is a question of decision-making and ultimately of choice, something which has become politically embedded in the ways that policy-makers attempt to change travel behaviour (Thaler and Sunstein, 2008).

Second, ideas of individual choice do, however, only represent a particular view of how researchers have attempted to understand the 'decisions' we make about travel. An emergent theme within sociology, and one that has spread into related disciplines such as geography, is the role of what Shove (2003) calls broader 'social practice' as a different framing of how we understand individual travel behaviours. This perspective highlights the historically embedded social trends that shape what we see as behaviours. For example, one could look at the individual decision to travel by car to and from work. Yet this could also be considered a form of social practice termed 'commuting'. In doing so, we begin to understand more about the role that underlying social trends and processes play in shaping shared behaviours. Commuting is an activity that has emerged from a series of processes which have necessitated and made desirable the daily journey to and from work. As Chapter 2 outlines, this was initially a necessity through the over-population of urban centres, but commuting was rapidly packaged up with the affordances supposedly offered by suburban living and the separation of the 'dirt' of the city and the wholesomeness of the countryside. Indeed, the invention and marketing of the motor car only accelerated the spread of suburbs and the embedding of a commuter culture that goes largely unquestioned. As such, the social practice of commuting has both a history and a social context; it is a phenomenon that reflects more than individual choices, but rather crystallises the embedded habits and lifestyle formations that we take for granted.

Finally, therefore, we come to the broader context for understanding transport and mobility, which relates to the infrastructures and technologies that underpin the way we travel. Here we refer to the planned landscape in which we live and the technologies that make our world knowable. A considerable amount of research on transport and mobility has been developed from the perspective of isolated human agency (Shaw and Hesse, 2010); that is to say it has regarded travel choices as something autonomous and often unrelated to place. In contrast, we argue that mobility is intricately related to the characteristics of the places in which people live, work, take leisure time and ultimately they seek to belong. This means re-establishing an intellectual connection between the study of movement and the characteristics of place and what promotes dwelling (Glaser and Shapiro, 2001). In so doing, we wish to challenge researchers to consider how our contemporary living arrangements promote or suppress the need to travel. One example we will use in this book that illustrates our argument is the way in which an intricate relationship has been established between private motor transport and the contemporary suburban pattern of settlements that has emerged during the twentieth century. As Kunstler (1994) has argued, there is a politics to suburban development that reflects the affordances given to the motor industry that largely erased other forms of urban development based around inter-urban railways, streetcars and metro systems. Much of this has to do not only with the practical benefits provided by the car; it is also about a symbolic association of the motor car with the political and social values of the dominant classes within nations. In the USA, for example, the automobile culture has become bound up with the so-called American dream of a single family dwelling located on a sacred plot of land, with a double garage, yard, two cars and access to the amenities offered by the freeways that lead to the malls that sustain consumer culture (Baldassare, 1992). Indeed, as Rajan (2006, p. 113) has highlighted, this form of living, entirely dependent on the motor car, has become part of the liberal agenda that holds up individual freedoms above all else:

> Automobility, on its part, has become the (literally) concrete articulation of liberal society's promise to its citizens that they can freely exercise certain everyday choices: where they want to live and toil, when they wish to travel and how far they want to go.
>
> (Rajan, 2006, p. 113)

In this way, car ownership seems to render place an almost unnecessary burden, given that travel is so easy. Kunstler (1994) has argued that the result of this kind of living arrangement produces places that lack character and ultimately are dysfunctional. He has also argued that such places are unsustainable; they are entirely reliant on cheap oil, without which it would not be possible for inhabitants to sustain the long commutes, trips to shopping malls and the necessary drives for maintaining everyday social relations. Accordingly, the contemporary living arrangements shared by so many citizens in the developed world today raises questions about the future viability of suburban development designed around the motor car. In an age of anthropogenic climate change and the potential for oil price rises and supply instability, there are major questions surrounding the future of the suburb as a desirable, affordable and sustainable settlement pattern (Dennis and Urry, 2009).

Quite apart from these pragmatic concerns, there are also questions about the social consequences of planning places that are stitched together by the motor car. Here we encounter the intellectual controversies surrounding what constitutes and sustains a 'sense of community'. As authors such as Baldassare (1992) have argued, auto-suburbs have resulted in ways of living that have placed family relationships under significant stress. Indeed, they have often resulted in the decline of city and town centres, in favour of monotonous tracts of identical dwellings, with few amenities and a dearth of public transport links. The city has often become the victim of out-of-town shopping centres, with centres being the preserve of downmarket shops and drunken night-time economies.

In short, there has been a gargantuan failure of planning that commenced before the Second World War and which was accelerated after its ending as planners scrambled to rebuild or redevelop city and town centres fit for a brave new age of Modernism. The traumas of war therefore blighted many urban landscapes as pre-war icons of civic pride were demolished in favour of functional, linear and progressive architecture. In the city of Exeter, south-west England, where much of the city centre was destroyed in German bombing raids during the Second World War, a vision of urban redevelopment was created around the car, which necessitated the removal of many fine examples of Victorian architecture (Sharp, 1946).

It is without doubt that such visions were designed with complete integrity and thought of as great improvements for the people who lived and worked in these places. Yet what they have created is nothing short of a mobility nightmare; the presumption that the main mode of transport would be the motor car, affording personal freedom and the necessity to build more and more infrastructure to support its growing needs, from brutalist car parks, inaccessible highways and spaces of exclusion for pedestrians and cyclists, to an overall architectural approach that was built not for people but for machines. In essence, we have created a built environment that locks us into one way of moving around that has come to represent our values, values that speak of transience rather than belonging, movement rather than dwelling and superficiality rather than substance.

What (auto)mobility says about ourselves

If our current reliance on the motor car and its embedding within the way we have planned our towns and cities says something about how we value place (or not), then it also reflects the wider framing of our values. Ultimately, the way we live and travel also says something about who we are, what we value and what drives us as individuals. We live in an age where success is often associated with affluence that is accumulated at the individual level and which is to be demonstrated by both the accumulation of material possessions and the accruing of experiences, which themselves have an exchange value. Put another way, our society is one dominated by the consumption of goods, services and experiences, which must be constantly renewed, invigorated and innovated to persuade us to replace and renew and thus promote economic growth. Such a growth model is one that has come to dominate modern life in the developed world and is one that is largely aspired to by those in developing nations. Recognising the links between the consumer-based economy and the spatiality of mobility is critical if we are to understand the full implications of trying to understand how new and potentially lower carbon forms of mobility might be envisioned for the future. And in our view, the state of our current economic praxis means that such visions are likely to remain just that – fantasies for the future rather than realistic options for radically reducing our reliance on fossil fuels.

Accordingly, recognising mobility as a reflection of wider economic circumstances requires us to consider two sets of praxis. On the one hand, we are aware that our economy is driven by the consumption of material goods and that much of what we deem 'the good life' is determined by our accumulation of everything from clothing and cosmetics to cars and houses. The accumulation of these goods requires immense investment in mobility both from ourselves and from the systems of provision that enable these goods to be delivered and sold at ever increasing speeds. From the human perspective, it is only since the Second World War that shopping has become a recreational activity of choice for the majority of populations in Western nations. Yet this accumulation of material goods is something that requires significant investment in mobility, through regular trips to supermarkets, shopping malls and out-of-town retail centres and garden stores. The very accumulation of things, therefore, has become an act of mobility in its own right.

However, an often untold story is how these things, our materials of desire, come to us through the systems of provision that have been created to service an economy based on consuming more and more products. There is often an unseen transport geography to the distribution of goods that has constructed a global network of material mobility that is reliant on air, ship, rail and road distribution, much of which is finely tuned to deliver goods in specific time slots. Indeed, the revolution afforded by the construction of railways in the second half of the nineteenth century, enabling fish from the Highlands to be served in London within a matter of hours, has been up-scaled to the transference of goods across thousands of miles within hours. Accordingly, what we wear, the food we eat and the consumables we buy are artefacts of a form of globalised hyper-mobility of materials, which relies largely on cheap oil to fuel this distancing of the consumer from producer. As such, it is worth considering, the next time you open a packet of salad or even a pint of milk, how far the product has travelled (sometimes several times along the same route) before it reaches the supermarket.

Finally, of course, there is the economy of the non-material. Just as we accumulate value in today's economy from the ownership of things, we also gain a great deal from collecting and sharing our experiences. In what Pine and Gilmore (1999) refer to as the 'experience economy', late modernity is defined by the value of exchange not only in physical products, but in the status to be gained from using experiences as a form of social exchange. And once again, this is where contemporary mobility reflects the wider economy. Tourism and travel, as this book will attest, have shifted from being highly regulated, routinised and standardised forms of practice to ones where the seeking out and evocation of individual experiences has come to mark the shift from a Fordist model of tourism production (mass tourism, holiday camps, package holidays) to a post-Fordist framework, in which bespoke, individualised, tailored and niche travel have come to dominate the sector. In so doing, tourists themselves become innovators and sensation seekers, willing to invest time and money in the seeking out of new luminal experiences that can be framed (literally, using photography) and packaged as unique experiences of great commodity value when sharing such experiences with friends, family and work colleagues. Accordingly, Larsen *et al.* (2007) have argued that the value associated with tourism and travel for many individuals is something that has spilled over into everyday life and is used as a mode of exchange and social networking far beyond the end of a holiday, but within a range of life settings. Indeed, if the rise in so-called gap-year travel for school leavers is examined, it is evident that this kind of mobility acts as a social networking device for many years after its completion, through the sharing of common experiences, cultural assumptions and the symbolic value associated with having visited and 'experienced' particular places. In essence, this kind of mobility reflects a desire to belong to a certain economic class, able to demonstrate its power through its collection and consumption of exotic destinations.

Mobility therefore tells us much about who we are and what we value. And what we value seems to be less and less about where we live, but rather how we can gain access to the world outside and, when we are at home, how we can surround ourselves with the material 'comforts' of living the so-called good life. This of course tells us much about our economy; an economy based on material accumulation at the individual level, about personal goal setting, personal achievement, freedom of choice and a form of liberal democracy that often sets us apart from, rather than entwined us with, those around us.

Accordingly, we want to use this book partly as a device for considering the ways in which mobility reflects these values and whether an immobile lifestyle, based around a reconnection with place, could offer us something different. Without question, this would necessitate a change in values and a shift in economic practices (Baritz, 1989). But this may be required in any case, if the aspirational targets to reduce anthropogenic climate change emissions are to be met (Crompton and Thøgersen, 2009) and we are to come to terms with the long-term challenge of declining stocks of cheaply available oil (Heinberg, 2004). It may well be the case, therefore, that even if we don't want to give up on our obsession with mobility, it may be forced upon us.

However, notwithstanding these instrumentalist concerns, it is also the case that reconnecting with place has value in its own right. Anyone who has walked down a British high street in a town or city that was bombed during the Second World War will recognise that additional and considerable damage was heaped upon the urban landscape long after the bombers had left, by the futuristic principles of

Modernist planning and the car-based utopian thinking of so-called progressives. These places are often drab, single-use, large-scale and socially sterile places of coming and going, but they are not places of dwelling. They do not encourage sitting, stopping or appreciation. Instead, they are a nod to a functionalist approach that views the town or city as a place to 'do businesses', but not one that has character, intrinsic value or is representative of history and topography. In short, we have often planned places that are non-places, artefacts of sterility and dysfunction.

In this book, we call for a reconnection to place as a way of reimagining good places to live and a vision for towns and cities that is not solely based on economic growth at any cost, but one that is grounded in making places for dwelling and for inclusion. This means radical change in the ways that we consider how our towns and cities look, how we choose to live in them and how we develop them to make way for rising populations. We argue that this ought to be done in such a way that promotes dwelling and a pedestrian-based society that rebalances the (important) role of the private car with other forms of mobility, and which fosters belonging.

Such a manifesto for change is controversial, because it challenges assumptions about the value of mobility for a growing economy, but the changes we argue for are hardly controversial, because they appeal to what most people would imagine are good principles, notably that towns and cities should represent the history and topography of their foundation, that they should offer a diversity of accommodation and mixed land uses, that they could have safe and walkable streets with neighbourhood facilities, and that the quality of their environment was such that people weren't keen to get in their car and go somewhere else to spend leisure time.

What we are calling for, moreover, is not a wholesale stop to mobility, but a rebalancing of priorities, away from the necessity to move and to do so for long distances, but the right to choose – to choose between a high-quality dwelling place and the alternatives. And in so doing, we want to encourage a critical engagement in how people consider what it means to lead a good life. Perhaps one component of the good life is moving around less and dwelling a little more.

Structure of the book

This book aims to provide the context, theoretical setting and practical background for understanding the contemporary challenges for transport and mobility. Part I will continue to introduce the historical context of our modern patterns of mobility through examining the birth of mechanised transport during the industrial revolution of the nineteenth century, focusing on both the UK and North America. In Chapter 2 we will explore the rapid expansion of cities and towns through the lens of rapidly changing transport technologies and will outline the factors that led to initial suburban growth and then the rapid expansion of such suburbs through the widespread ownership of private motor transport. Chapter 3 will then set the policy context for suburban growth through examining the motivations behind the modern planning system and the ways in which this has shifted to become a system that has largely operated a 'predict and provide' road building programme to service the auto-mobile society.

Part II (Chapters 4 to 6) will then set transport and mobility into a theoretical and conceptual context. We will examine the history, development and principles underlying conventional transport geography and mobilities research, before exploring how geographers have utilised these different approaches in the setting of both daily travel and tourism and leisure travel. In doing so, we aim to illustrate how different theoretical framings and methodological approaches offer alternatives for understanding transport and mobility, representing the diversity of perspectives in the social sciences.

Finally, Part III aims to demonstrate how the challenges of anthropogenic climate change and resource security lead us to reconsider how we value mobility in the modern world. In Chapter 7 we outline the ways in which sustainable mobility has been framed and institutionalised and then in Chapters 8 and 9 we explore the duel dilemmas of how to effect behavioural change alongside the promotion of more sustainable places to live – places that will encourage dwelling and ultimately reduce both our need to travel and the necessity for the products and services which rely so heavily on mobility.

Ultimately, what we aim to argue in this book is that current framings of transport and mobility are representative of an economy based on consuming things and many different places. This pattern of consumption is unlikely to last for much longer. It is probable that we will have to learn to live with less mobility, fewer things, less exotic travel. Our argument is that by thinking about our connection to place, to home, to dwelling, we can rebalance mobility with immobility and realise the benefits and pleasures of a place called home.

2 The 'long mobile century'

From streetcar suburbs to auto-mobility

Introduction

Without doubt, the twentieth century has witnessed a revolution in mobility that has changed the lives of billions of people, not only in terms of how they personally experience travel, but in the products, services and technologies that afford experiences, which were once imaginaries fed by a few travel writers who made difficult and dangerous journeys around the globe. To use the temporal signifier of a century seems appropriate in this instance because it is the twentieth century that witnessed the major innovations in mobility which have resulted in the car-dominated culture that we live with today. Indeed, it was the twentieth century that saw the first powered flight that ultimately ushered in the jet age of air travel, which enables billions of people annually to travel long distances.

Yet the twentieth century is only part of the story and the origins of what we have come to know today as mobility are set back in the mid-nineteenth century and are still being outworked today in the twenty-first century through evolving forms of technological innovation, living arrangements, commuting, leisure pursuits and tourism. Accordingly, it is fitting that in describing the period since mass mechanised transport emerged, we recognise how this has developed as a long but clear narrative of enhanced mobility that has fundamentally changed our relationship to space and place, and which has ensured that the time we spend on being mobile is unrecognisable to that which would have been the case 200 years ago.

In this chapter, we aim to situate contemporary debates on transport and mobility within this important historical context. We start by examining the development of mechanised transport in the mid-twentieth century, focusing on both the revolutionary steam locomotive for long-distance travel and the impacts of technologies such as the streetcar and underground railway for commuting within settlements. We track what we term 'sprawl' through the late nineteenth century and early twentieth century, in which commuter railways and electric streetcars provided not only the ability to move longer distances, but in many ways determined the geography of the places where they operated. We also examine the ways in which the notion of the suburb began to emerge as a highly stylised, aspirational and commodified space through the development of particular elite developments.

In analysing this first period of sprawl, largely afforded by transport networks, we also examine how the geography developed by public transport was progressively transformed into a private transport network, one made possible by the invention of the motor car and in particular the mass production of vehicles, pioneered by Ford's production line. What emerged from this was a period of suburban development that would far outstrip anything possible with the initial streetcar networks, which was the birth of the auto-suburb, enabled largely by state-funded road building programmes and, in the case of the USA, the construction of the interstate freeway system. In analysing this second phase of sprawl, we will explore the growth in suburban cultures that has evolved around the Utopian fantasy of what Kunstler (1994) has termed the 'little log cabin in the woods' and in so doing we will explore the identity politics of suburban cultures.

Our aim in focusing on both the streetcar suburb and the auto-suburb is to explore how mobility is intricately related to how our contemporary living spaces are configured and the ways it is implicated in how we relate to place. Mobility, we argue, says something about who we are as individuals and our broader social context. Moreover, as Chapter 3 will outline, mobility has been a goal that has been enthusiastically supported by politicians because of the economic benefits provided by the movement of people, goods and services.

Sprawl I: the streetcar suburb

An examination of Greenwood's 1830 map of London (Bath Spa University, 2017) demonstrates that just before the development of the steam railway network, the city of London and its associated suburbs were already a patchwork of narrow streets that would have been crowded with horse-drawn traffic (Figure 2.1). Nonetheless, most workers in London at the time would have been able to walk to work and would have lived relatively close to their site of employment. Yet with the growth in industrialisation, huge pressures during the mid-nineteenth century were being placed on urban centres and this initially spurned an entrepreneurial approach to transport.

The conventional way to get around a city like London in the first half of the nineteenth century was the Hackney Carriage, a licensed horse-drawn carriage, of which there were some 1,100 in 1805 in London (London Transport Museum, 2014). Yet the growth in the need for mass transit resulted in the development of new forms of transport, initially pioneered by George Shillibeer and his horse-drawn tram, which commenced service between the village of Paddington and the City of London in 1829 (Figure 2.2). The success of this initial service resulted in the lifting of the monopoly of Hackney Carriage drivers picking up fares from the roadside.

Horse-drawn bus travel during this period was hardly glamorous. As the London Transport Museum (2014, n.p.) highlights, one passenger in 1833 described using what became known as an omnibus 'decent clerks, fagged and harmless, going home for their tea. "Here we are," he wrote, "in all six and twenty sweating citizens, jammed, crammed and squeezed into each other like so many peas in a pod".'

Nonetheless, horse-drawn buses become popular out of necessity and by 1839 there were 620 licensed buses in London. The challenge of the horse-drawn

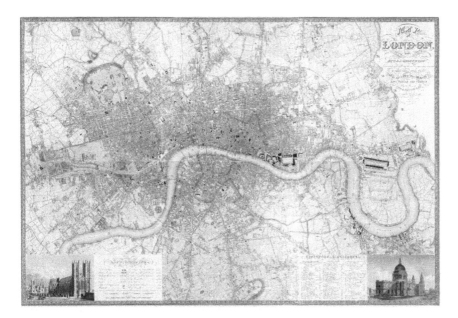

Figure 2.1. Greenwood's 1830 Map of London.
Source: Motco Enterprises Limited, ref: www.motco.com.

Figure 2.2. An 1829 Shillibeer Omnibus.
Source: © TfL from the London Transport Museum collection.

omnibus was that it was a relatively expensive mode of transport compared to its speed and comfort, not to mention the number of horses required to pull a bus of 20 to 25 people. Accordingly, in the 1850s London developed a system of horse-drawn trams that ran on metal rails embedded in the street. Introduced by George Francis Train, these trams could be pulled by two horses and the reduced friction on the road provided by rails meant that they were faster and more efficient (Figure 2.3).

The horse as the major powerhouse of urban transport in cities like London led to numerous challenges for city authorities. Pragmatically, the 90,000 or so horses that were used in public transport during the latter part of the nineteenth century in London led to around 1,000 tonnes of dung being dumped on the streets daily (London Transport Museum, 2014). The sheer space required for stabling and care for the animals was immense, including the servicing of horse shoes, feeding and veterinary care. Overall, streets became highly congested with carriages and horse buses and into this congestion came another factor that would increase traffic – the suburban and national steam railway system that was being developed to link London with outlying settlements and of course other cities.

The first commercial passenger railway to be built in the UK was the 1825 Stockton and Darlington Railway on Teesside, but this revolutionary venture soon spurned a raft of major inter-city projects that would result in the construction of major termini, initially to the north of London and towards the west at Paddington. London's first mainline terminal was opened in 1837 at Euston and from here, London didn't look back, with major stations being added largely in their current form throughout the 1860s with their attendant railway hotels.

Figure 2.3. An 1888 horse-drawn tram.

Source: © TfL from the London Transport Museum collection.

London soon became a centre for transport and throughout the Victorian era, the city was able to become a workplace for many more people living in adjacent towns, as suburban lines and stopping services were added alongside the main-line inter-city routes. This added pressure on the city's infrastructure and meant that getting to and from mainline stations to places of work, such as the City of London, became a slow and unpleasant process. Such was the challenge of nav-igating the city's streets in morning and evening rush hours that a revolutionary solution was proposed and eventually approved in the early 1860s: the construc-tion of an underground steam railway, to link the existing mainline termini and the city.

London's underground opened in 1863 and linked Paddington with the city via other mainline stations. As an engineering project, it was immense, but largely carried out on time and to budget, with few accidents. The route was constructed through what has been termed the 'cut and cover' method, where streets along the route were excavated, lines laid, brick tunnel roofs built and the road surface relaid. Travellers along this part of London Underground's route network today can still view original brickwork and get a sense of what a mid-nineteenth century station would have looked like in places such as Baker Street station (Figure 2.4).

The underground was a steam-based railway and as such, regular ventilation was required from the tunnels laid just metres below street level. Accordingly, despite the vast improvements in speed and efficiency provided by this new rail-way, it was a largely unpleasant and dirty experience – one that was by no means universally enjoyed by commuters. Nonetheless, it set the scene for the practice of commuting that we know and experience today. It meant that people could live

Figure 2.4. Baker Street station in 1863.
Source: © TfL from the London Transport Museum collection.

further away from the city and their place of work and take a mainline train into a London terminal and then transfer onto an underground train for their final leg of the journey to work.

The underground created a culture of commuting that is familiar to us all, but it also created a new domesticity. As Jackson (2003) has highlighted, the development of railway suburbs was a key feature of the period from 1850 to 1914, when railways provided the only opportunity to commute even quite short distances. Initially, the costs associated with commuting and the suburbs developed alongside new railway lines were high and suburban development was largely for middle-class households. Yet significant lobbying by the London County Council resulted in 1903 in the publication of a 'quarterly return' (Jackson, 2003) in which stations offering cheap return travel for commuters was made available for manual workers (see Halliday, 2013).

Nonetheless, the development of the Metropolitan Railway (the initial underground line) did spur on significant development of middle-class housing that spread out from central London in a ribbon development and which became immortalised in the advertising associated with new homes that afforded space, fresh air and gardens for food growing. A glance at the extensive collection held by the London Transport Museum (2014) highlights the ways in which this new living arrangement was marketed (Jackson, 2006). 'Metroland' provided an escape to the country from the dirt, pollution and overcrowding of the city. It afforded new ways of living that had all of the benefits of modern life – indoor bathrooms, gas stoves, large living spaces, gardens for both growing produce and cultivating flowers, and crucially, these were all located within walking distance of a rail or underground station (Figure 2.5) (Levinson, 2008).

Just as Metroland was developing a unique geography in the hinterland of London, a far more expansive and nationwide project of suburbanisation was occurring in North America. Stilgoe (1988) and Thompson (1982) have both provided accounts of early suburbanisation in the USA, which followed the same compelling logic of that in the UK, namely the twin motivations of finding space for living outside of overcrowded settlements, alongside a need to move away from the dirt and squalor of the city. Yet suburbanisation in the USA and Canada took on a very different form because of the ways in which land was owned and how new lands, especially on the western frontier of expansion, were sold off to developers. Booker (1977) and Kellett (1969) have demonstrated how the British system of railway development was a contested and lengthy process of negotiation (sometimes unsuccessful) between railway companies and landowners. Indeed, the historic land ownership model of the UK meant that suburban developments occurred sporadically along railway routes, much according to the preference for the landowner to sell his property. By contrast, the USA and Canada, at least on the western fringes, was a blank canvas. Baldassare (1986) documents how lands were mapped and allocated according to the well-known US grid system, with lots being sub-divided as required. Because of the control exercised over land allocation by the USt government, an orderly system of land allocation and transport network provision could be developed that did not have to follow a

Figure 2.5. Homes in Metroland.

Source: © TfL from the London Transport Museum collection.

historically rooted pattern of land ownership and land thus became available in a logical ordering away from city centres in sequences as railway lines and streetcar lines opened up for business.

A useful example of this process is provided by Hayes (2005) in his discussion of how suburban development occurred in the new city of Vancouver, British Columbia, Canada. Much of what is now metropolitan Vancouver was initially endowed by the Canadian government to the Canadian Pacific Railroad (CPR), who were constructing a transatlantic line that would terminate in British Columbia. The original assumption was that the line would end at New Westminster, a bustling colonial port on the Fraser River. However, the affordances provided by the aptly named deepwater Coal Harbour (Burrard Inlet) soon made it a good contender for the line's terminus. Eventually the small hamlet of Gastown was settled upon (consisting before the 1880s largely of timber merchants and saw mills). Upon the announcement that this would be the site for the new CPR terminus, the city of Vancouver was founded in 1886, the year operations started on the completed line.

Vancouver at this time was a small city but it grew rapidly, in large part due to the way in which building lots were sold off during the first decades of the twentieth century. The CPR would hold auctions at which land dealers, or in the case of one sought-after suburb, individual families, could buy lots for building. Entrepreneurial streetcar operators soon exploited the opening up of dense forest lands for land speculation and initial suburban development was plotted along lots that were within easy walking distance of stations (Figure 2.6).

To this extent, a map of many US cities during the first part of the twentieth century would be representative of the intrinsic links between streetcar lines and speculative development either side. This meant that large tracts of land between lines, out of walking distance, went underdeveloped for residential use. But what this did mean was that there was a largely suburban-based system of living that was dependent on public transit, with relatively little road traffic and the consequent development of suburban centres within neighbourhoods based on walking.

Within Vancouver, the streetcar legacy is one that still lives on in the retail landscape of today. As Hayes (2005) has highlighted, identifiable neighbourhoods with particular characteristics grew out of suburban developments that were based around streetcar stations. For example, suburbs like Arbutus Village, Dunbar, Kitsilano and Kerrisdale became largely self-contained districts that were linked to other parts of Vancouver by streetcar and inter-urban railway, but enabled citizens to shop, eat and do business locally as well.

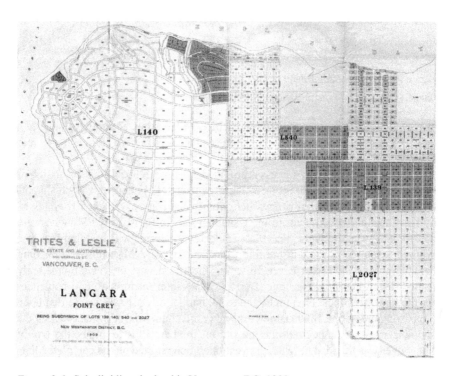

Figure 2.6. Sub-dividing the land in Vancouver, BC, 1909.

Source: Reproduced with the kind permission of Derek Hayes.

Along with the development of real estate, often outside of city boundaries, came the attendant advertisements about the affordances provided by suburban living. These were based on, although hardly in line with, the first Utopian visions of suburbia provided by the late-nineteenth-century attempts to construct an urban nature in the form of suburbs like Riverside in Chicago (Baldassare, 1986). These exclusive developments had much more to do with securing a pleasure ground for the rich and famous of the late nineteenth century and have much in common with contemporary attempts to produce gated communities in places such as Hollywood. These were highly stylised environments, built and landscaped to a neo-colonial specification that attempted to blend civilised living with the wilderness so emotive in US culture. Moreover, they were environments to be coveted by the masses, aiming for self-improvement and a better quality of life:

> These early suburbs [...] captured the general public's imagination and were often spoken of in almost utopian terms by urban planners, politicians, and private developers. Their presumed benefits included urban decongestion, lower residential densities, greater separation from the city's business district and, importantly, home ownership.
>
> (Baldassare, 1992, p. 476)

To understand, particularly in a North American context, the role of suburbs in a collective psyche, we have to explore the emotional roots of the suburb (Baritz, 1989). Suburban aspirations were not only linked to a functional dislike of the city as a space of squalor but were also connected to the counterpoint of urban dwelling, which was the great American wilderness. As Baritz (1989) notes, the wilderness mentality has been a powerful part of US mythology for hundreds of years and was wrapped up with notions of expansionism and the lone explorer owning a log cabin in the woods and living off the affordances of nature (Kunstler, 1994). The suburb is a kind of civilised translation, in US culture, of this dream through a rejection of the urban and a tentative embracing of nature through the trappings of an individual house and garden, often adorned with the rustic trappings of wicket fences and clapperboard claddings (Figure 2.7). What this means is that for most Americans, there is a desire to have the independence and space afforded by the suburb, while at the same time living in the human-created comforts of a domestic setting that is anything but connected to nature (Hayden, 2003).

As the second half of this chapter will attest, the construction of an 'American Dream' surrounding suburban life has developed much more into a set of practices and characteristics that must be attained for entry into the middle class. Indeed, Baritz has argued that:

> America's spirit and tone, its historical mythology and official aspirations, political bent, educational arrangements, the centrality of business enterprise, as well as the dreams of the vast majority [...] derive from the psychology of the great imperial middle.
>
> (Baritz, 1989, p. 1)

Figure 2.7. A stereotypical American suburban 'tract' home.
Source: BrendelSignature at the English language Wikipedia.

It is to this 'great imperial middle' that we now turn. For it was this newly enriched, suburban middle class that the purveyors of the motor car marketed their new product, a product that would revolutionise suburban living beyond all recognition.

Sprawl II: the age of the automobile

In describing the private motor car, Kunstler (1994) has written that there has never been an invention in history that has so successfully freed humans from what he terms the 'bondage of place'. He is of course paraphrasing, but it is without question that there has never been an invention that has so remorselessly enabled the individual to be 'free' from the limitations of walkability, public transport and ultimately from interacting with other people. In the private motor car we find a space of privacy and freedom, enabling us to go where we want, when we wish to and with whom we please. In this way, Rajan (2006) has argued that the car represents the very literal articulation of freedom in a liberal society, a theme we shall return to later in this section as we consider the political out-workings of the automobile age.

The car's entry onto the world stage of transport technologies was enabled in large part not by the successful invention of a proven technology, because early motor cars were often slow, unreliable and expensive to run. They were clearly the preserve of the wealthy and it was not really until the simplicities of Henry Ford's production line that mass production of vehicles commenced on a scale that could make the car a viable proposition for those on regular incomes (Brinkley, 2003). In a country like the UK, this meant the production, during the 1920s of smaller, more efficient and slightly more affordable cars, although car production and sales did not in any way emulate that of the USA.

To give an overview of the trends in car ownership, a House of Commons (1999) report noted that in 1926 there were 1,715,000 cars on UK roads, an increase from just 8,000 in 1900. A set of traffic counts at 467 roadside points in 1935 revealed that an average of 11 'mechanically propelled' vehicles passed each point every hour. In 1954 this had risen to 159. By 1998, road traffic had increased by 500 per cent compared to 1955. Accordingly, apart from a sharp decrease in vehicle registrations during the Second World War, car ownership has continued to rise steadily. In the USA, we see a more rapid increase in car ownership during the early years of car production, not least because of Henry Ford's production technique.

The impacts of the motor car on personal mobility can be viewed as both direct and indirect in nature (Jeekel, 2013). Kunstler (1994) has evocatively argued that the car, in direct terms, represents the first piece of technology that has enabled the individual to be liberated from the 'bondage of space' and as such represented a fundamental change in the psychology of how we related to space. No longer did people need to consult a timetable, walk to a tram stop, find their fare or worry about the last train. Here was a piece of technology that enabled complete freedom to roam. As such, Kunstler (1994) has argued that it satisfied a deeply held aspiration amongst many Americans, rooted in national mythology, to 'push west; that is, to go where no one else has gone, to go under one's own steam and to partake in appreciating some of the wilderness that their forefathers had conquered a century before'. As such, the motor car provided the perfect technology to live out the American dream of personal freedom and liberation from the constraints of space.

It is without doubt that the car did and still does afford many personal freedoms. The passing of the driving test is still a much coveted event for teenagers eager to be relieved of the constraints presented by parental oversight. Indeed, the motor car has offered freedom of mobility to many who could, until recently, not have access to public transportation. This has been enriching to family life, social life and has enabled access to places that could not have been reached before, or at best would have required significant planning and expense. For example, it would have been near impossible for a family to enjoy a day out on one of the peaks of the Lake District in England before the motor car arrived; today a simple and low cost trip is available to the majority of people who can travel at their own convenience. The car therefore represents for us a considerable personal liberation and democratisation of mobility in the context of what went before it.

Yet we must also be aware that the direct impacts and benefits of the car as an individual artefact of mobility have also come with consequences that are indirect in nature. In this way, we must temper our superficial appreciation of the motor car by considering its deeper impacts, both on our personal mobility and also in the context of what the car has meant for our towns, cities and our ways of life.

To turn initially to the car as the artefact of personal mobility, it is certainly the case that the car does offer us more mobility than any other technology has in the past. But we must also recognise that the car also leads us to behave, and to think, in particular ways about ourselves. Rajan (2006) has argued that the car perfectly represents the articulation of personal mobility and liberty in a neo-liberal society, where the value of individual freedom is placed above all else.

In framing auto-mobility in this way, Rajan contends that the car has facilitated the process of moves towards an individualistic society, which has become such an important political project in Western society in the twentieth and early twenty-first centuries. The car, then, embodies a set of wider aspirations to uphold the individual (Giddens, 1991; Rose and Miller, 1992) as the most important element in society, neatly set within broadly defined routes and boundaries:

> Its constitutive visual image is one of dignified convoys of individual cars, vehicles whose solitary drivers can remain separated from each other as they collectively pursue private goals on public highways. As such, this picture captures the salient features of cars in a post-Enlightenment order: the experience of driving, identified by the quiet pleasures of the open road, speed, power and personal control, neatly complements the functionality of covering distance, managing time and maintaining certain forms of individuation.
>
> (Rajan, 2006, p. 113)

As we shall see in Chapter 5 and also Chapter 6, studies of auto-mobility have delved deeply into the ways in which people have developed particular mobility practices and how the materiality of the car has become, in some cases, an extension of the human body as part of an identity-forming process. However, for the purposes of this chapter, it is sufficient to note that the car is not only about giving us personal freedom to move through physical space, it is something that cultivates individualism, which itself is part of a wider underpinning project of political change in Western societies.

The car has therefore revolutionised our personal, individual lives and has enabled us to exercise choice about everything from when to go shopping, to making a spontaneous decision to have a day out at the beach. Yet it has also led to some indirect changes in our physical and social environment that have had profound impacts on how we conduct our daily lives. First and foremost, the motor car enabled the project of suburbanisation to continue well beyond the natural limits placed upon it by the constraints of public transport infrastructure. For example, as we discussed earlier in this chapter, suburban settlements during the nineteenth and early twentieth centuries tended to follow the line of metropolitan railways or streetcars, either as ribbon developments or settlements clustered around lines. There was only a short tolerable distance that people could be expected to walk to reach a station. However, the emergence of the motor car provided the means to both infill between the gaps left in streetcar suburban development, as well as permitting entirely new suburbs to be constructed, beyond any influence from public transport infrastructure.

However, the emergence of auto-suburbs, certainly as we know them today, was not by any means inevitable. Accordingly, a second indirect impact of the motor car was the way in which suburban development took place and the impacts this had on cities. The context for major changes in our towns and cities was undoubtedly the emergence and popularity of Modernism as an architectural influence, which originated in the 1930's push towards the adoption of art deco creations in pre-war European and North American cities. Grindrod (2013) and Kunstler (1994, 1998) have provided both popular and highly engaging accounts of the impacts of Modernist thinking on our culture and built environments, but in essence the message of Modernism was about creating places to live, work and move through that were aligned to the technological advances of the day and which embodied functionality, rationality, linearity and logic. Kunstler (1994) has argued that Modernism's popularity can be related to the adoption by the Nazis of neoclassical forms of architecture and that the adoption of Modernist principles for rebuilding by post-war governments reflected a need to move on from Hitler's obsession with the classical. Indeed, the trauma of six years of war, coming only a few decades after the First World War, must have cemented the need to wipe away the old and to start afresh. Indeed, in many European cities, and in the UK in particular, there was such a level of destruction and damage that a near blank canvas was provided on which to plot new settlements, built for the twentieth century and which could harness the technology of the future.

What this future was to look like can still be seen in many British towns and cities today, although thankfully much of the ambitiousness of post-war planners was only to make it into prospectuses for city rebuilding, published as grand documents of hope for a better future. A useful illustration of Modernist principles in this era of frenetic activity after the Second World War can be seen by examining the plan to rebuild the city of Exeter, south-west England. Published by the Architectural Press in 1946, Thomas Sharp's *Exeter Phoenix* was an elaborate prospectus for the rebuilding on Exeter after the bombing that had left much of the city badly damaged or destroyed in May 1942. Exeter was part of Hitler's so-called Baedeker raids, ordered after the British bombed Lubeck in Germany in 1941. Hitler is said to have commanded his generals to find the most attractive cities in England and destroy them. Exeter was one of these cities, which was incorrectly regarded as being a safe location during the war, having received evacuees from London to escape the blitz there.

Sharp's prospectus for Exeter was immense and profound. It was founded on a set of principles, which guided many an urban rebuilding scheme at the time:

- Road-based communications;
- Remodelling the city for the car;
- Separation of people and traffic;
- Modern, iconic public buildings;
- Single-use zoning for residential, leisure and work areas;
- Overall, an efficient, mobile city.

Perhaps the most important of these was the way in which Sharp believed that the city had to be entirely remodelled to accommodate the motor car. He took inspiration from North America, as evidenced in the way in which he discusses his plan for an inner ring road 'freeway':

> ... this new central by-pass will be quite free from any kinds of junctions and will have no building fronting upon it. It will be what in America is sometimes called a 'freeway' [...] In it Exeter will possess perhaps the most perfect internal by-pass of any city in England.
>
> (Sharp, 1946, p. 76)

Sharp's plan was to create an American style freeway to allow traffic free reign around the city. This would have necessitated the destruction of the 1838 iron bridge, the iconic Rougemont railway hotel, a considerable number of (now highly desirable) terraced houses and the introduction of fast traffic to many peaceful areas of the location immediately outside of the city walls.

Within the city centre, Sharp advocated the separation of people and traffic through various devices, the most innovative of which was the underpass:

> The site of the Rougemont Hotel is an almost ideal location for the city's bus station [...] it will permit of a complete and highly desirable separation of vehicular and passenger access [...] Passenger access will be from Queen Street (by easy ramps), from the new Iron Bridge, and also directly from the Queen Street Railway Station by means of a short passage under Queen Street – this latter providing a much desired arrangement, but one not yet achieved in any English city.
>
> (Sharp, 1946, p. 79)

Although an architectural innovation at the time, it is interesting that the age of the underpass has thankfully been short-lived, especially as it privileges cars over people in their access to natural daylight, relegating pedestrians to grim and dimly lit subterranean walkways.

Sharp also embraced the values of Modernism through his dislike of what he thought of as dysfunctional buildings, such as the city's museum and art gallery (now the second most important architectural gem in Exeter after the Cathedral). In his prospectus, he described the museum and art gallery as 'an architectural horror: it is also extremely gloomy, and none too well-suited for its purpose' (Sharp, 1946, p. 98).

Finally, Sharp adopted the model of land-use planning that is still prevalent in the USA today, which was to argue for single-use zoning throughout the city. This was a quite understandable approach at the time, as many cities in the UK and North America were plagued by industrial outlets in or near city cores, which made life unpleasant and no doubt reduced life expectancies. Yet Sharp failed to recognise the waning of British industry and the shift to what we would come to know as the service-based economy. In this way, he advocated a model of development that drained the city centre of residential dwellings and necessitated travel into the city for work, shopping and leisure. Indeed, as we shall see later in this chapter, the net result was simply to abandon city centres as cheap land outside city boundaries was developed for out-of-town retail parks, endowed with large car parks and coffee shops for light entertainment.

Modernism therefore flourished as a result of the technological innovation represented by the car and we have never really escaped its legacy, for it fostered a way of thinking about places that has built in the need to travel to fulfil the most basic routines of daily life. Development after development has followed the model of equating suburban living and a motor-based mobility with the 'good life'. This was neatly encapsulated in a 1946 Central Office of Information (COI) film produced by the British government to show the benefits of building and living in new towns. In contrast to the dirt, over-crowding and backwardness of the city, living in a new town would be a new beginning:

> Our town was going to be a good place to work in, and a grand place to live in, with plenty of open spaces; parks, and playing fields where people could enjoy them, flower gardens, and of course there'd have to be an attractive town centre too, with plenty of room for folks to meet. Good shops, a posh theatre, cinemas, a concert hall, and a civic centre.
>
> (COI, 1948)

The effect of all of this rebuilding and creation of new towns on green fields was varied, in terms of its physical impacts. Without doubt, some development was aesthetically sympathetic and did manage to follow topography and form in the landscape. But much of the Modernist legacy in the British landscape today is brutal in nature. Indeed, across the Atlantic the impacts are even more profound (Figures 2.8 and 2.9). Huge swathes of city cores have been sacrificed to the motor car, while freeways have eaten up tracts of land outside of cities, where alongside malls and motels all the trappings of US auto-consumption are placed in sub-divisions of isolated housing, linked only by wide and menacing freeways.

This physical legacy, facilitated by the car and guided by the (often brutalist) principles of Modernism has led to a third and perhaps most fundamental indirect impact of auto-mobility. This is the impact that an automobile culture has on our daily lives. Later in this chapter we discuss the cultural implications of suburban living, but notwithstanding this aspect, the automobile culture in which we live has wrought massive changes in both the geographies of everyday life and the governance of the places in which we live.

As we shall see in Chapters 7, 8 and 9, we are now attempting to undo many of the changes that witnessed the rise and rise of the motor car as the travel mode of choice and this has largely come about because of a backlash against the dominance of motor transport. Banister (2008) has highlighted how a generation of transport planners have viewed development that privileges accessibility by motor car as the preferred option. This has led to profound changes in the ways we live our daily lives. To list but a few examples, we might point to the dominance of

Figure 2.8. The Alaskan Way viaduct, which cuts off downtown Seattle from the seafront.
Source: Waqcku at English Wikipedia.

out-of-town supermarkets and retail spaces, the development of trading estates on the outskirts of towns and cities, the construction of large shopping malls along-side motorways and freeways, the homogenisation of high streets and main streets as town centre shops try to compete with out-of-town retailers, and the planning of new settlements that are largely dependent on their residents owning cars. All of these changes mean that we travel further, for longer, to do many of the things previous generations did nearer to home. Be in no doubt therefore, the car certainly guides our lives.

The influence of the car has changed our built environment, perhaps forever. Yet during the transition from a walking and public-transit society to one based on long commutes in cars, there has certainly been great upheaval and intense political unrest. Two examples serve well to explore what the very different outcomes of such political geographies of mobility can be. In the first instance, we can look at the momentous legacy of perhaps the most powerful road engineer of all time: Robert Moses, who ran the Triborough Bridge and Port Authority in New York through the 1940s to the 1960s. Moses is regarded as being largely responsible for the revolution in New York freeways that resulted in a series of schemes that criss-crossed the city and dislocated tens of thousands of inhabitants and forever tore through communities all in an effort to meet the apparent demand for freeway

Figure 2.9. A clover-leaf freeway interchange in Michigan, USA.

Source: Michigan Department of Transportation.

access (Caro, 1974). The result of Moses' work is a city that has been literally shaped by the freeway and its attendant feeder roads and the physicality of grid-irons and tarmac. Caro (1974) noted, in the era immediately after Moses stepped down from his power base in charge, how he was responsible for nearly every one of New York's iconic freeways. As Kunstler (1994) notes, Moses' authority spent $4.5 billion during his chairmanship of the Bridge and Tunnel Authority, all of it on roads and not on public transport, leaving a legacy that still lasts today, with a still circuitous journey to JFK International Airport. As a result, New York is a city that is still recovering from the infliction of so much disruption and lack of investment in public transport and community development. Moreover, for most of America, it set the scene for a long narrative of freeway building. As Caro (1974, p. 10) states: 'More significant than what Moses built was when he built it. That was how he put his mark on all the cities of America'.

Yet this approach to freeway construction and city planning more generally was not adopted wholesale. One notable example is provided by Hayes' (2005) illustration of what he terms the 'freeway fight' in Vancouver, BC. Following the trend in the rest of North America, proposals were brought forward in the early 1960s for a network of freeways to facilitate traffic movement from north to south, across the downtown area, involving a new crossing of the Burrard Inlet (to complement the pre-war Lions Gate Bridge). As Hayes (2005) notes, it was widely assumed that Vancouver would have a downtown freeway system and the only remaining question was where these should be located. Numerous studies ensued, in the first instance being led by a US freeway consultancy company. This involved, in part, the construction of a freeway on land owned by the Canadian Pacific Railway along Vancouver's now iconic waterfront, which was the backdrop to the 2010 Winter Olympics and the entry point for many cruise passengers. Yet despite the many plans and studies, downtown Vancouver today is absent of freeways. As Hayes notes, Vancouver's final decision not to adopt a downtown freeway system can be attributed to several factors. First, by the late 1960s there was growing evidence from US cities that 'universally told a story of blight and neighbourhood deterioration in the vicinity of swathes cut by elevated urban freeways' (Hayes, 2005, p. 156).

Second, there was concern amongst publics that corporate interests were driving the plans to build freeways. Third, the opposition formed to fight freeway construction spoke largely with a single voice and represented a broad range of interests. Moreover, public opposition was able to influence the City Council and Hayes (2005) provides an entertaining account of how a public meeting marked a shift in official opinion:

> … it was Campbell [the freeway advocate Mayor] himself who provided the final but unwitting blow with a memorable piece of verbiage: 'The new crossing', he was reported as saying, 'is on the verge of being sabotaged by Marxists, Communists, pinkos, lefwingers and hamburgers'. The latter he defined for the *Vancouver Sun* as 'persons without university degrees'.
>
> (2005, p. 157)

In the event, a mass petition persuaded the Canadian federal government to withdraw their support and elections resulted in an administration in favour of mass transit over freeway construction.

The case of Vancouver is important, but it is an isolated example amongst a predominant trend for mass freeway constructions within and between cities, as evidenced by the early completion of America's Interstate freeway network. In the UK today, Conservative government policy is for more road building and fewer planning restrictions. As we shall see in the following chapter, the philosophy that 'roads drive economic growth' is very much still with us. Yet roads, vehicles and all of the associated paraphernalia of the automobile society do have their impacts, direct and indirect. Most importantly, they have left us with a legacy of suburban living that has meant that we have developed new ways of living with our residential environment, which is neither in the city nor the country, but somewhere in between. It is to the cultural implications of this way of living that we now turn.

Cultures of suburbia

The places we have come to know as suburbs, which for the majority are home, have become immortalised in many different ways and have become both part of a cultural aspiration and an academically critical discourse (Huq, 2013). Indeed, the popularisation of the suburb has emerged as an artefact of cultural inquiry, as the BBC's (2014) exploration of *Sounds of the Suburbs* illustrates through music and film. Accordingly, to fully understand the growth of suburban areas and their attraction, we need to consider what Baritz (1989) has argued is the meaning of success for the middle classes. In his view, success is packaged as a set of material aspirations comprised of a spacious detached dwelling, a yard, double garage, two cars and the latest consumer appliances. In hit TV programmes like *Happy Days* (Miller-Boyett Productions, 1984), the construction of family life, gender relations and relationships between people and materials provided a set of aspirations to work towards that equated a certain level of material wealth with success and well-being. Consider, for example, the stereotypical and highly gendered nature of product advertisements from the 1950s, which ascribed gender roles to the male *breadwinner* and the female *housewife* (Youtube, 2008a). Such advertisements presented the car as an absolute necessity that afforded complete freedom and the flexibility to live a life unbounded by the restrictions of space.

Suburban cultures (Conley and McLaren, 2009; Gans, 1967; Jackson, 1985; Price, 2000; Silverstone, 1997) have been the focus of considerable attention by sociologists (Baldassare, 1986, 1992) and studies like that of Gans (1967) have revealed much about the practices and prejudices of suburban living, as his in-depth study of Levittown in the USA testified. More than anything else, the ability of blue-collar workers – in Europe those we have termed the working class – to own a property was a major change to the aspirations of the majority of society. With these changes came shifts in family relations, living arrangements and social practices. Perhaps most notably was the shift from living arrangements that were dense and inter-linked in nature to a focus on the nuclear family. In countries such as the UK, the clearance of large areas of slums during the immediate post-Second World War period meant the dismantling of communities that had been largely inter-dependent through the shared need to use resources such as water and heating. In the new housing developments on the edges of major cities, these previously densely populated communities became separated and rehoused into individualised units, each with brick and wooden boundaries.

Such changes, as Giddens (1991) has noted, have reflected the general shift in post-war society towards a focus on individualism and the reliance of family units on the consumption of goods to satisfy the needs previously met partly by co-operation and kinship. Yet it must also be recognised that in the immediate post-war period, radical improvements in material quality of life and the affordances of new homes with potable water, heating and gas appliances offered a marked improvement in quality of life that may, at the time, have taken precedence over the less tangible benefits of supposed community cohesion. Accordingly, movement to the new suburbs was treated with excitement as individuals began to discover the benefits of living in enhanced comfort, albeit now a bus or car ride away from shops and services.

This new age of suburbia has been popularised in many different ways (Huq, 2013). In the UK, situation comedies and feature films such as *Mutiny On the Buses* provided excellent examples of the lifestyles and technologies of working class Britons during the 1960s and 1970s (Hammer Films, 1972). Such programmes highlighted not only the nature of family relationships but the strains placed on such relationships in an economy that was changing rapidly and becoming more reliant on private transport as the main mode of travel for the majority. Indeed, *On the Buses* also provided an image-scape of the changing relationship between the built environment and personal mobility. The opening credits of the film is a tremendous tour of early 1970's suburbia, epitomised by the bus company's name: 'Town and District'. Background scenes of newly built housing estates, suburban shopping centres and the spread of low density, low-rise houses capture the essence of lives becoming mobilised by an extended living environment. In a contemporary setting, from the USA, any number of situation comedies and dramas provide clichéd views of the suburb. Take, for example, the lifestyle practices exhibited by Charlie, Alan and Jake, inhabitants of a beachfront property in California in the iconic and highly popular *Two and a Half Men* (Warner Brothers, 2015). Here is suburbia at an extreme, almost taking us back to the original grandeur and privacy of Riverside in Chicago. In the situation comedy, Charlie's beachfront home is also occupied by his brother Alan and Alan's son Jake, who are living there after Alan's marriage has broken up. Each scene in the sitcom has specific and repetitive place settings: the large home, the motor car, the diner, the mall, the bar and Charlie's exorbitant therapist. These place settings are completely set in isolation, with no clear geography linking them, except by means of the car. In each place, the aesthetic of a stereotypical suburban life is reinforced, from the characterless restaurant, to the over-sized refrigerator, to the seemingly temporary and superficial nature of the therapist's consultation room. No one is ever seen going for a walk, a bike ride, taking a transit or simply sitting in a park admiring the view. Without the beachfront location, the sitcom could be anywhere in North America, so it seems.

As with popular TV, film has documented the growth of auto-suburbs and affords opportunities to gaze on the lives of those who inhabit the sprawl of conurbations like Los Angeles. Take for example the iconography of the *Fast and Furious* movies, of which the first is a quintessential study of the North American sexualisation of the motor car, alongside a narration of apparently dysfunctional suburban life. The film depicts a Los Angeles hijack gang, who regularly attack trucks on desert highways. Undercover cop Brian O'Connor is tasked with infiltrating the gang, in whose work he becomes embroiled. In the first instance, *The Fast and the Furious* (Universal Productions, 2001) is concerned with the core instincts of private car ownership as it has been constructed in Western society: speed, flexibility, adoration, competitiveness, sexualisation and power over others. However, the film also reveals much about the darker side of suburbia, in which a criminal class can play on the vulnerabilities of others. It is seen as a place in which law-breaking can be viable, given its sheer size, anonymity and monotony.

The underlying discourse of such films, with which we are so familiar, is one that depicts the suburb as dysfunctional and socially destructive. As Clapson (2003) has argued, suburban life has often been signified by particular gender, race and class relations that have often led to both prescribed forms of behaviour and segregation. Indeed, Clapson (2003) further argues that suburbs in the immediate post-war era in the USA were defined by a moral encoding of suburban life that was, for many, highly restrictive and repressive.

Much of the focus in such critical readings of suburbia has been the seeking out of a lifestyle and attendant practices that accord with an ideological worldview of what the 'good life' should be like (Baritz, 1989). In this way:

> The suburban ideal is about finding a homogenous community of like-minded people [...] This valuation of suburbia for a particular form of family life has thus provided some of the primary ideological underpinnings for the exclusionary actions of individuals moving to suburbia, as well as for the various government policies and business strategies that have built suburbia.
>
> (Miller, 1995, p. 993)

Suburbs have therefore been cast as places that are exclusionary, prescriptive about gender roles, individualised and which lead to family breakdown and low levels of personal well-being (Baldassare, 1992; Dagger, 2003). Indeed, for Kunstler (1994) what is particularly worrisome is the dis-connection from place that occurs in suburban environments; blocks of uniform retail units, large parking lots, identical wooden-framed dwellings and large shopping malls all provide a landscape of sameness that lacks identity and discourages belonging.

Such images and depictions of suburbia are just that: they are popular. However, there is a danger that in either valorising or condemning suburbia as one form of built environment, we omit to recognise that suburbia is about the cultures which exist there (Nicoladies and Wiese, 2006). While we may recognise particular forms of suburban development and note its monotony on one scale, Archer *et al.* (2015) remind us that too many of the critiques of suburban living have made generalisations that do not stand up to scrutiny. This is because:

> Suburbia is so varied that it is impossible to define it in any one way. It is not a single place. It is not even a singular kind of place. Suburbia is a complex and richly textured physical and social fabric, multiple terrains of varied and vital places, practices, and identities.
>
> (Archer *et al.*, 2015, p. vii)

As such, Archer *et al.* argue for a reading of suburbia that builds on the seminal works of authors such as Jackson (1985) and Thompson (1982) who have traced the histories of social change in suburbia, alongside Wiese's (2005) exploration of suburbanisation in particular ethnic contexts. In an impressive edited collection, Archer *et al.* (2015) brings together researchers who illustrate the complexity and dynamism of suburbs, arguing fourfold that, first, suburbs have enabled and have encouraged the mobilisation of particular minority groups; second, unlike the popular stereotype, suburbs have attracted considerable memorialisation and sense of place amongst their inhabitants; third, suburbs have fostered new ways of being and gathering, from the culture of the Asian Mall to the literal gathering of mega-churches; finally, suburbs continue to be defined by diversity in building styles, and resultant domestic practices, from the emergence of Garage Rock to the outdoor kitchen.

Conclusion

An evaluation of the impacts of over 150 years of suburbanisation and the impacts wrought first by steam mass transport and then the internal combustion engine cannot be neatly summarised in a few paragraphs. However, we can discern a number of key trends that characterise the most momentous period of human progress, which has unquestionably been largely defined by our ability to move people, goods, services and ideas with great speed and efficiency.

First, we might pose a simple but often unposed question: did it have to be like this? There is, without doubt, some inevitability about the way in which we think mobility has out-turned. In other words, there was always going to be a move from public and mass transport to systems that privileged private and individual movement. We should at least question this assumption and ask why some nations and cultures in the developed world are much more comfortable with mass transit than others. As this chapter has illustrated, we cannot explore how mobility has developed without first understanding the culture and psychology of the places we are studying. In North America, the early adoption of liberal principles and the upholding of individualism have undoubtedly had an impact on how the new technology of steam and then electric power codified suburban development. However, it is also the case that while the early quarter-acre lots and wooden homes of newly deforested land did represent a built-in individualism, they were originally planned by and with the streetcar and inter-urban railways in mind. The arterial routes of cities like Vancouver, BC, were designed specifically to allow residents to be within walking distance of streetcar lines. So it was not inevitable that when the motor car was invented, this would all simply be torn up (as it was). Rather, the great weight of political and economic interest fell behind the motor car as a driver for economic growth, employment and the much coveted aspiration of 'progress'. The resultant ripping up of tram lines was hailed as a step towards freedom for millions who coveted the car; yet only decades later, we are desperately trying to reinstall mass transit to both reduce congestion and to revitalise the urban centres that the car forgot.

Second, the increasing amount of mobility experienced by populations has not only led to greater personal opportunities for travel, but has changed the very way in which life is lived. The link between mobility and lifestyle practices is key to note here. Very few of us would necessarily make a connection between the development of motor transport on the one hand and the types of homes we live in. Yet the emergence of mass-produced homes, from the mock Tudor 1930s semi to 1960s imitation Scandinavian terraces with in-built car ports testify to the impact the car has had on how we live. Indeed, the way we shop, eat and take leisure time have all changed since the motor car has become a must-have product, making 'old ways' of consuming much more problematic. In short, mobility has shaped all of our lives, but not in simple and linear ways; rather, mobility and lifestyle are mutually (re)producing effects continuously.

Third, the emergence of a largely hyper-mobile society has occurred over a relatively short timescale. This has meant that new practices of mobility have been adopted rapidly and have changed within generations. Such changes are often momentous but also difficult to negotiate and cope with. For example, it is well within living memory that the first bulldozers began to clear land ready for a new London Airport in the late 1940s. At that time, and during the airport's initial phase of development, it would probably have seemed inconceivable that any further runway capacity would be needed. Indeed, for those householders who purchased 1930s homes bordering the Great Test Road nearby, it would have been unthinkable that a jet carrying upwards of 500 passengers would fly over their property every 90 seconds during the day. This is why transport and mobility are such intensively contested issues: because they often pit our aspirations for one kind of life (in the domestic sphere) against our aspirations for mobility. Yet it is the very possibility for mobility that has resulted in these forms of domestic life: a suburban house, with a garage and no need to live in the urban core. Accordingly, in the following chapter we examine how a country such as the UK has been shaped by the discourse of mobility and the power of what we have come to know as 'predict and provide' for planning our transport needs.

3   Predict and provide

Transport planning in an age of climate change

Introduction

This chapter plots the policy landscape that has led to the current state of transport planning, using the United Kingdom of Great Britain and Northern Ireland (UK) as a case-study example. In particular, it focuses on the ways in which an intricate but often hidden relationship has existed between land-use planning and mobility.

Much space, particularly in urban areas, has been given over to support car-based mobility, including road infrastructure and sites for refuelling and parking. The increasing emphasis on car-based mobility led to the construction of motorways and ring roads. For the sustainable transport modes of walking and cycling, an increase in road infrastructure has often presented barriers for trips to the centre of towns and cities. The car has the advantage over more sustainable forms of surface travel in that it offers the freedom to travel in any direction over any distance. Public transport tends to be confined to fixed routes while walking and cycling tend to be confined to short trips.

Transport planning is concerned with the way that generalised aspirations, such as sustainable mobility, are translated into physical outcomes on the ground (Headicar, 2009, p. 277). These outcomes vary across the various spatial scales, from local and regional sustainable transport elements such as the provision of walking, cycling and public transport facilities, up to the national provision of road and rail infrastructure. At the international level, along with planning for aviation and maritime-based transport, there is the increasing emphasis on environmental issues such as climate change.

This chapter examines the development of transport planning in the UK, and then how environmental issues have become an increasingly important component within the discipline, before bringing these elements together on the implications for the geographies of transport and mobility.

Development of transport planning in the UK

Taking a chronological approach, this section considers how these initial attempts to create better places were seen alongside the continued and largely unplanned expansion of urban centres and the encroachment into 'natural' landscapes, alongside the development on unsustainable personal mobility based around the motor car.

Motor-car mass production may have begun in the early parts of the twentieth century, but it was only really following the Second World War that motor-car mobility increased rapidly. The increased post-war prosperity, alongside supportive British politics, enabled car use to grow over the 1950s and 1960s. Car-based mobility emerged from the privileged few to the wider UK population. The actual, and more importantly predicted, growth in car use heralded an era of road-building, which encompassed motorway construction and made planning for roads in urban areas a priority. Transport planning entered the so-called 'predict and provide' era, providing car-based opportunities in response to the insatiable demand to own and use such vehicles.

Tensions between land use and the motor car are therefore not recent, but date back to the 1970s and before. In the second half of the twentieth century car-based mobility increased rapidly in the UK. In the 1960s there was a realisation, as documented in the influential 'Traffic in towns' Buchanan report (Ministry of Transport, 1963), that unlimited use of the motor car could not reasonably be sustained, particularly in urban areas.

However, an era of increased reliance on the car-based mobility in developed countries continued, such that there became a recognised term of 'automobile dependency' (Newman and Kenworthy, 1989). This influential work of Newman and Kenworthy ranked 32 cities in terms of motor-car dependence. The highest ranking and most motor car dependent tended to be the US and Australian cities. Conversely, the lowest tended to be European and Asian cities. Interestingly, therefore, UK cities fall somewhere in the middle, between the sprawling cites of the USA and Australia, but more car dependent than most of their counterparts on mainland Europe.

There was a dramatic increase in travel over the 1980s and 1990s, that despite many developed countries towards the end of the first decade in the twenty-first century facing recession, the mid-term and long-term trend is for travel demand to increase. Road traffic has increased by 85 per cent since 1980 until 2008 (Department for Transport, 2009), although much of this rise was during the 1980s. Car-use growth has been hand-in-hand with the rise in disposable incomes. In addition, although the real costs of motoring have remained at a similar rate, public transport costs have increased. The average time that people spend travelling has hardly changed over the last 50 years or so (around 1 hour per day), but nowadays individuals are travelling further.

In response to this demand, a policy to provide road building and associated infrastructure to accommodate the motor car continued between the 1960s and 1990s. During this time period the policy focus changed from traditional 'predict and provide' road building as solutions to the problems of increasing traffic, to more efficient 'travel demand management' of the UK transport system. The concept of travel demand management was introduced in the 1990s with a realisation that demand for the motor car outstrips road supply, irrespective of the amount of road building undertaken. Towards the end of the 1990s, instead of building new roads, the UK government changed focus to maintain existing roads and to manage the road network more efficiently to improve reliability. This approach to the problems posed by the strategic trunk network encouraged multi-modal studies (Department of the Environment, Transport and the Regions, 2000a), whereby for the first time sustainable modes of transport, including walking and cycling, were accounted for in new road schemes.

The policy change towards travel demand management was strongly linked to the 1998 Integrated Transport White Paper 'Transport: A new deal' (Department of the Environment, Transport and the Regions, 1998), which represented a change in government transport policy, heralding the so-called 'Integrated Transport Strategy'. An underlying theme of the White Paper was to reduce the need for travel for individuals, achieved by encouraging a modal shift to sustainable transport modes such as public transport and non-motorised modes, and reducing car dependency. Policy goals included providing choice between the different transport modes, ensuring efficient integration between the different modes of transport. People may have a freedom of choice between transport modes, but there are often imposed constraints such as an ability to own and use a motor car and the availability of public transport services.

Following the Integrated Transport White Paper, a Ten Year Plan was produced in the year 2000 (Department of the Environment, Transport and the Regions, 2000b) with a focus on transport delivery. It was arguably concerned more with investment in roads and railways than with sustainable mobility. It represented a policy shift back towards the motor car and road building, and away from the environmental concerns, although both aspects were still evident in the transport policy. Despite the many UK transport policies generated around this time, such as the Integrated Transport White Paper, there has been much scepticism as to whether the sustainable transport policy agenda can deliver (e.g. Docherty and Shaw, 2003). The principles behind the Integrated Transport White Paper have been widely welcomed and agreed upon by most commentators, but the primary problem associated with this transport policy has been the ability to implement (Goodwin, 1999).

The Integrated Transport White Paper (Department of the Environment, Transport and the Regions, 1998) attempted to tackle the 'twin pillars' of air pollution and congestion. Begg and Gray (2004) argue that a combination of public dissatisfaction with progress in transport, political shocks (primarily the national fuel duty protests) and institutional change have led to a policy shift away from integrated transport. With more focus since the Integrated Transport White Paper on the problems of congestion (at the expense of air pollution) and the provision of transport infrastructure, they assert that the 'marriage' between transport and the environmental policy is over.

In the subsequent decade there has been an uneasy relationship between transport planning and environmental policy, central to the concept of sustainable mobility. The next section charts the development of the environmental debate within transport planning up to the current time, with climate change now playing a more prominent role.

Development of the environmental debate within transport planning

The relationship between transport and the environment can be set in the context of broader sustainable development. The start point for this discussion is a definition and description of sustainable development. Implicit in the sustainable development concept, to meet 'the needs of the present without compromising the ability of future generations to meet their own needs' (World Commission on Environment and Development, 1987, p.41), is that the way that humans behave today should not harm prospects for future generations. From this definition of sustainable development, within the 1987 Brundtland report, and through subsequent global UN Conferences on Environment and Development, such as the Rio Summit in 1992, sustainable development has evolved from a marginal concern to the political mainstream.

Sustainable development does have a core tension between economic and environmental goals, although the original concept is broader, incorporating other issues such as population increase, poverty, technology and social organisation. There is a sustainable development contradiction: environmentalists claim that it is cover for continuing to destroy the natural world, while economists claim that there is excessive concern about depletion of the natural resources (Dresner, 2002, p. 2). Aviation, for instance, may be not environmentally sustainable, but it is economically and socially sustainable (Upham, 2003). The tensions between the economic and environmental tensions relating to aviation can be shown in the, albeit simplified, view for (economic) and against (environmental) proposals for development of airports in London and the south-east of England. There is a spatial planning aspect here, too, with limited land available for large-scale transport developments such as airports. Sustainable development has enabled most countries in the world to discuss and agree progress, although a tension remains between developed countries, such as the UK, which have more of a concern for environmental protection alongside economic growth, in comparison with developing countries which naturally have a clear focus on economic development.

In terms of policy implementation of sustainable development, the UK government's response in 1994 to the global stabilisation targets set at the Rio Summit, was to develop a UK Sustainable Development Strategy (Department of the Environment, 1994). The next strategy, in 1999 (Department of the Environment, Transport and the Regions, 1999) had four over-arching principles: economic growth, protection of the environment, social concerns and natural resources. It included 15 measurable indicators of progress, but in order of sustainable development success were the economic-based indicators, followed by the social-based indicators and finally the environmental indicators. The third, and most recent, UK Sustainable Development Strategy (Department for Environment, Food and Rural Affairs, 2005) had a more explicit focus on environmental limits. Climate change has also had an increasing role within the concept of sustainable development, as shown by its more prominent role within this strategy, as one of the four agreed priorities.

In terms of the role of transport within sustainable development, this can be demonstrated from the list of 68 sustainable development indicators generated by the UK government (Department for Environment, Food and Rural Affairs, 2011). Of the indicators, the following eight indicators are directly transport-related:

- Aviation and shipping emissions (Greenhouse gas emissions from UK-based international aviation and shipping fuel bunkers at airports and ports);
- Road transport (CO_2, NOx, PM10 emissions from all road transport);
- Private cars (CO_2 emissions);
- Road freight (Heavy Goods Vehicle (HGV) CO_2 emissions);
- Mobility (Number of trips by walking / cycling and public transport / taxi);
- Getting to school (Children walking / cycling to school);
- Accessibility (Differences in access with and without car);
- Road accidents (Number killed or seriously injured).

These transport indicators cover a wider range of mobility aspects and it could be debated whether they fully represent sustainable mobility. Such a list tends to focus on the easily measurable aspects, and transport also occurs indirectly through other indicators, such as carbon dioxide emissions by an end user across a range of industry sectors (including transport).

This leads to the discussion of the attributes that constitute a sustainable transport system. Early on in the debate, Black (2000) stated the following criteria: sufficient fuel for the future; minimal pollution from that fuel; minimal fatalities and injuries from motor vehicle accidents; and manageable congestion. However, the concept of sustainable transport has increasingly needed to take more account of climate change. This influences attempts to categorise transport modes, which can be simply put as a reduction in travel that has high carbon emissions, such as by aviation and by motorised transport. Conversely there will be an increase in travel with no or low carbon emissions, by the so-called sustainable transport modes of walking, cycling and public transport. Public transport may usually emit less pollution and causes less congestion per occupant than the motor car, but the levels vary by vehicle type and occupancy level.

The increasing role of climate change in transport planning

There has always been a globally changing climate, but in recent times it has been possible to statistically link changes, associated to increased levels of key greenhouse gas emissions, to human activity. This is causing an increase in the earth's average temperature, the so-called 'global warming' effect. In addition to meteorological evidence, the impacts of climate change can be demonstrated from physical geography consequences such as ice melt in the Arctic.

Most scientists would agree that there is a climate change phenomenon (i.e. linked to human activity) and questions remain over the scale of the process. Much of the uncertainty relates to the difficulty with the prediction elements of climate models, in turn due to the complex weather systems that are involved.

The Intergovernmental Panel on Climate Change (IPCC) has the role to assess the scientific basis of risk of human-induced climate change, based on peer reviewed and published scientific / technical literature. The IPCC Fifth Assessment Report (IPCC, 2014) states that it is extremely likely that human activity is the dominant cause behind global warming, and surface temperature will rise by the end of the twenty-first century according to all modelling scenarios tested (some generated an increase of over 2°C). There will also be an increasing level of unpredictable and extreme weather patterns, including heat-waves and tropical storm intensity.

In terms of climate change, transport is a major contributor to greenhouse and pollutant emissions, and transport tends to be the only industry sector where emissions have been increasing. This is particularly pertinent in aviation, where particular fears relate to the recent and forecast growth in air travel even though in proportion terms the level of global emissions is a very small proportion. The levels of emissions from aviation growth means that the sector will take an increasingly significant proportion of any carbon budget (Anderson *et al.*, 2007).

The Kyoto Protocol, originally agreed in 1997 and subsequently ratified as the Kyoto Treaty, committed industrialised countries to overall 5.2 per cent reduction in emissions of main greenhouse gases from 1990 to 2008–2012. As part of the agreement, the UK had a target to reduce greenhouse gas emissions to 12.5 per cent below 1990 levels by 2008–2012. The UK met this target, partly due to the manufacturing decline and changes in the energy sector 'dash for gas', when many UK energy companies built new gas power stations, which are a cleaner and cheaper alternative to coal. There was a major concern that carbon dioxide emissions from international aviation and shipping were not accounted for in the Kyoto Treaty, and that there has been an absence of an internationally agreed methodology for allocating these emissions at the national level.

The Climate Change Bill (Department for Environment, Food and Rural Affairs, 2008a) has a commitment for a UK reduction of 80 per cent in CO_2 emissions by 2050 based on 1990 levels, and an interim target of at least 34 per cent by 2020. The UK has been the first country to make a legally binding commitment to cut greenhouse gases and the targets are very challenging. The UK policy response on the economic response to climate change, through the Stern Review (Stern *et al.*, 2006), commissioned by the UK government, has called for the aviation industry and air passengers to cover their external costs of air travel (cost of climate change).

There are two responses to the challenge of climate change, mitigation and adaptation. Climate change mitigation refers to 'anthropogenic (human activity) intervention to reduce sources or enhance the sinks (natural sources) of greenhouse gases' (IPCC, 2001). The environmental mitigation measures for surface transport can be summarised as (from Chapman, 2007): encouragement of a modal shift away from (sole occupancy) motorised transport, 'stick' and 'carrot' transport policy measures, technological improvements, driving improvements and new fuel developments.

Mitigation measures can be classified as 'sticks' and 'carrots' (Stradling *et al.*, 2000). 'Sticks' force individuals away from a mode (e.g. increasing cost and decreasing availability), while 'carrots' encourage individuals in transport choices, enticing them towards a mode (e.g. public transport facilities). It is important to get a balance between 'sticks' and 'carrots' measures. It is beneficial to sell the mitigation measures, promote the carrots ahead of sticks, but sticks are known to be more effective. A balanced approach is therefore required, for example that money from the stick of congestion charges is 'transparently' ring-fenced to the carrot of transport improvements. There has been an increasing trend for transport measures to be directly linked to the carbon emissions from vehicles. For instance, the level that vehicle owners pay in taxation or charge has been applied to UK vehicle tax (vehicle excise duty) since 2009.

Climate change adaptation refers to 'adjustments in natural or human systems in response to actual or expected climatic stimuli or their effects' (IPCC, 2001). To recap, climate change will decrease some weather events, such as the amount of snow and ice, but others, such as higher temperatures, will be more frequent. In addition, the weather is likely to be more severe. In terms of climate change adaptation, there are two ways in which transport can be adapted in response to the changing weather: through infrastructure approaches and social impact. In terms of the road and rail network infrastructure, they are affected by issues such as flooding, landslides, rail-buckling and user visibility. Under the Climate Change Act, major UK organisations responsible for key aspects of national infrastructure have to explain how they will adapt and cope if the climate alters as forecast. This includes transport organisations such as the Highways Agency, Network Rail, and Transport for London, and, more importantly in an aviation context, large airports such as Heathrow Airport.

It is also of interest to consider the broader social impacts in response to a changing climate, and how transport mobility would change in response to climate change effects (e.g. people may not make an air-travel journey or use an alternative transport mode). There would also be a change in background social and economic relationships as a consequence of climate change, such as population change and an ageing population.

It is acknowledged that alongside climate change impacts there are continuing broader environmental impacts from the use of transport. The act of travelling produces pollutants that can have local air-quality impacts as well as global climate change implications. Noise from motor vehicles and aircraft is an environmental impact that can be a major concern for residents living near to major roads or airports. There are impacts from the life cycle aspects of transport, from the construction of infrastructure, such as road and railway development, to the disposal of scrapped vehicles and associated waste oil and tyres. Furthermore, the environmental impacts have an ecological dimension, as transport can impact upon the biodiversity of a particular area, for instance affected animal habitats. Finally, there are health concerns associated with some of the environmental impacts, including casualties from transport accidents, local pollutant impacts and noise disturbance.

This mix of environmental impacts means that the transport industry will often have to trade-off between them. For instance, air traffic controllers when determining aircraft take-off paths have to choose between local aircraft-noise implications for the surrounding communities and global carbon emissions. In an era of climate change, it will increasingly mean that these impacts will take priority.

Conclusion: contemporary mobility in an age of climate change

Following the introduction of the motor car there has been the challenge of how to respond to the insatiable individual demand for car-based mobility. The so-called 'predict and provide' approach, typified by road construction in the 1980s, was an initial response, but has since been replaced by attempts at integrated transport planning. However, as environmental concerns have grown over time, crystallised by the increasing realisation of the impacts of climate change, transport planning approaches have adapted accordingly. Banister (2002) conceptualises contemporary transport planning as 'understanding the process of change' (2002, p. 17), 'anticipating future growth' (2002) and 'evaluating options' (2002). This futuristic element is integral to both transport planning (e.g. through Local Transport Plans) and climate change (e.g. looking ahead to when the impacts will affect the way that we live).

Spatial relationships are important to understand in terms of policies that relate to transport planning and the environment. Climate change is a global phenomenon, whereas the transport planning and environmental policy responses are seen at the national, regional and local levels. Geography also comes to the fore in the development of sustainable, liveable places and associated transport mobility, and the underlying tension between maximising economic growth while minimising environmental concerns. We will return to these important issues in the third and final part of the book, before which we explore the conceptual underpinnings of transport and mobility and the ways in which these concepts have been deployed in research on everyday travel and tourism and leisure travel.

Part II

Approaches to transport and mobility

4 Transport geography and geographies of mobility

Introduction

The study of movement is perhaps one of the most quintessentially geographical themes of modern times, yet within the discipline of geography, transport and mobility have, until recently, received relatively little attention in the annals of mainstream publications. In one way, this seems surprising given the high policy relevance of transport and the technological advances that have enabled so many to travel long distances since the end of the Second World War. Indeed, as noted in Chapter 2, the transition to a hyper-mobile society has been so rapid and often so profound that it is odd that geographers have not considered fully the implications of personal mobilities on the lives and experiences of people who decades earlier might have just about expected to go on a British seaside holiday once a year. Notwithstanding the intellectual territorialisation that has meant that tourism research has often been located in business and management studies, rather than geography, our ambivalence as geographers to movement is puzzling. Yet seen from another perspective, the intellectual trajectory of geography can perhaps explain why transport geography has often been at the margins (although it is no longer, as we shall evidence). The difficulty from those interested in movement is that while human geography has undergone major changes in theoretical direction, these haven't, in large part, been reflected in the sub-discipline of transport geography. Indeed, transport geography has been highly successful at developing its own theoretical devices and methodologies from within and beyond its community, drawing on disciplines such as mathematics, statistics, engineering and psychology. As such, transport geography has acted as a bridge between disciplines, but in doing so it has not, until recently, been at the forefront of those subjects.

Transport geography does now, without doubt, have a far higher prominence within the broader discipline, for two unrelated but critical reasons. First, as we shall evidence in this chapter and Chapters 5 and 6, the so-called 'mobilities turn' in the social sciences has witnessed a revolution in both interest and intellectual innovation in how we understand movement as geographers, and how we can study it in all of its complexity. In part, the move towards a focus on mobilities has been motivated by a broader shift in social science to more critical readings of practices, subject and spaces (Cresswell, 2011). However, as formative scholars like Urry (2007, 2011) have noted, social science desperately needed new ways of describing and understanding the implications of what he has termed hyper-mobility. This has resulted in shifts not only in how we explore personal and collective mobility, but also in what we consider mobility to be, from virtual tourism to the movement of services, as well as our physical movement as human beings.

The mobilities turn, as we shall explore in this chapter, has not come without its critics and it has raised a range of challenges to longer lasting ideas about how to conceptualise and study movement. At times the debates in the literature have been acrimonious and drawn along over-simplified battle lines: scientific versus sociological, quantitative versus qualitative, and extensive versus intensive. Yet as Shaw and Hesse (2010) have noted, there is likely to be more gained from a coming together of ideas rather than from splitting apart. And that coming together may partly be driven by a second reason for a higher profile for transport and mobility in geography. With a renewed focus by nation states on getting value from higher education and research in particular, there has been a resurgence in what used to be called applied geography, that is, to use the jargon of the UK government (Department for Business, Energy and Industrial Strategy, 2016) generating research impact. Within the UK, as in other nations that have faced spending cuts for universities, there has been a shift to redistributing research funding towards projects that deliver tangible social, economic and environmental benefits. Indeed, it is now clear in the UK that transport is regarded as a key industry for promoting economic growth through infrastructure projects and resultant employment (Department for Transport, 2016c). Accordingly, interest in transport and mobility has risen steadily over recent years as researchers have been enabled to obtain research funding on topics that may previously have been regarded as too applied in nature. In some fields, this renewed interest has resulted in major expansion of research investment and expertise, most notably the major investments from both government departments and research councils into understanding behaviour change, which has been heralded as one of the ways in which the UK can cut its carbon emissions to reduce the impacts of anthropogenic climate change (Department for Transport, 2011).

The story of transport geography, and its association with the new entrant of mobilities research, is therefore important to appreciate, because it helps us to position individual studies in both an epistemological and methodological framework. This chapter starts this process by exploring the major characteristics of transport geography and mobilities research. In each case, the chapter will explore the theoretical and methodological frameworks that researchers have used to examine key issues, before providing a case study that is indicative of this kind of research. The key emphasis will be placed on interpreting these approaches for what they can offer, rather than what they lack. Indeed, our firm view is that a great many factors need to be considered when exploring what kind of approach should be taken to tackling a particular research problem, and this can often necessitate either a specific theory or method, or a mix. We end the chapter by highlighting what transport and mobility has to offer to geography as a discipline, which is an anchoring context for so many of the issues that are at the heart of geographical interest: the inter-relationships between space, place, time and scale.

Transport geography: overcoming space

Transport geography has a long and established reputation within geography as a sub-discipline, but it is certainly a field that has drawn not only from theory and practice in human geography, but has evolved facing towards a number of scientific and pseudo-scientific disciplines such as engineering and psychology (Keeling, 2007). In this section, we explore the major intellectual concerns and themes of transport geographers and we utilise research from some of the well-known studies (Hoyle and Knowles, 1992) to illustrate how transport geography has focused its attention on spatial relationships and trends at the macro and meso scale, often concentrating on regional case studies.

As Rodrigue *et al.* (2013, p. 1) have highlighted, despite the somewhat marginal nature of transport geography until recently, the study of transport is quintessentially geographical 'There would be no transportation without geography and there would be no geography without transportation.'

As such, transport geographers highlight the tension, or friction, that space presents as something to be overcome, in terms of physical distance, virtual distance, time, boundaries and so on. Accordingly, transport geography is interested in how human societies have sought to overcome space and to promote economic and social activity through movement. As Rodrigue *et al.* (2013) note, there are several key concepts that transport geographers use to explore these relationships, which are the building blocks of more sophisticated analyses. First, transport is a function of derived demand. In other words, transportation occurs for specific reasons in particular ways. Activities such as working, taking leisure time, holidaymaking, visiting friends and relatives, producing goods and food stuffs all generate demand for movement. This can be visualised as direct derived demand, which results in particular responses to activities (working necessitates commuting, holidaymaking necessitates air, rail and private car travel, producing goods and food stuffs requires freight transport through road, rail and sea transport). However, these activities can also be visualised as indirect derived demand (working, holidaymaking and leisure, alongside holidaymaking, all require attendant services, while producing goods and food stuffs require storage facilities), which ultimately is realised through the demand for energy. Accordingly, the study of derived demand requires us to consider transport as both part of an economic system and also embedded in social relations; it is therefore insufficient to simply analyse transport modes in isolation.

A second key element for transport geographers is the notion of distance. Rodrigue *et al.* (2013) note that transport geographers have been interested in three particular interpretations of distance:

- Euclidean: a measure of distance in particular units, such as miles, between an origin and a destination, along a straight line;
- Transport: a distance measure that reflects the actual distances travelled to complete a particular journey from origin to destination in relation to their time and costs, such as the range of travel modes and distances covered on a holiday (taxi to the station, train to the airport, flight to a foreign airport and shuttle bus to a resort, for example). In this case, distance is also a function of the very different times and costs involved in making a journey, some of which will be more efficient than others;
- Logistical: this final distance measure is concerned with incorporating transport distance into a more comprehensive analysis of what time and costs are involved in undertaking a particular journey. For example, in the case of holidaymaking, this could be related to the time and costs involved in booking rail and airline tickets, the costs of maintaining trains and aircraft, servicing these for the particular journey, maintaining stations and airports and the services that are provided in these transport hubs.

Once again, therefore, the focus transport geographers place on distance necessitates a wider appreciation of the economic and social processes that promote travel in particular ways and are associated with often hidden costs and time implications.

Three other key concepts that have dominated transport geography are worthy of note, before examining the ways in which these have been mobilised in particular research studies. First are the notions of access versus accessibility and distance versus time. As Rodrigue *et al.* (2013) note, access provides a simple measure of the physical availability to a transport system, but this does not equate to the ability to access the system. For example, a family living on a principal road that has a regular bus service and proximity to a rail station may not be able to access this network for a range of reasons, such as physical mobility impairment, low levels of household income or a disjuncture between their transport needs and the network's provision. Similarly, the assumption that shorter distances between origins and destinations will lead to shorter travel times must be questioned, since there may be a range of impediments. For example, speed restrictions will increase travel time, security procedures will increase waiting times at airports, and congestion will cause delays.

This first consideration leads us to consider a second, the nature of space–time relationships. It is without doubt that modern transport has radically changed such relationships, largely as a simple function of speed. Changes in this regard have come about largely in major leaps forward in technology and infrastructure. For example, in the late seventeenth century, a stagecoach journey from London to Exeter would take 8 days, but by 1844, when the Great Western Railway arrived, the journey time was just 5 hours, and it is now possible to fly from Exeter to London City airport in around 50 minutes. However, these improvements in raw travel time should not be seen in isolation from the underlying economics of transport. Through the rapid increase in passenger numbers afforded by new technologies, the economies of scale of operating at speed have ensured that more passengers can be conveyed at lower cost. Efficiencies in terminal facilities, and most recently the Internet, have lowered costs and, in terms of railways in the UK, have led to an upsurge in use.

Finally, transport geographers have become increasingly interested in the idea of networks. Networks refer to the pattern and shape of locations and links that form to transport people, goods and services. Importantly, such networks are not simply based on simple calculations of Euclidean distance or even transport distance, but are often historic artefacts of previous network development, economic decisions or even political ideology. As we evidenced in Chapter 2, the development of the US freeway system was without doubt an economic and political decision, as well as one founded on the simple principle of speed. Indeed, the contemporary construction of high-speed rail routes in Europe is arguably based not only on the demand for rail travel, but a political imperative to incorporate peripheral and economically depressed areas.

The study of transport geography

In line with a focus on the role of space, transport geographers have primarily been concerned with exploring two key dimensions (Knowles *et al.*, 2008). First, they have examined the changing nature of the geography of transport systems in the ways that such systems are developed, evolve and are used at a range of scales. This has clear links to policy and the role of different instruments of governance to determine both transport network development and the incentives and dis-incentives for promoting particular transport modes. Second, transport geographers have taken a keen interest in the relationship between transport and economic development, which has led to critical commentaries of the ways in which transport networks respond to and promote economic activity (e.g. Cullingworth and Nadin, 2006). Hoyle and Knowles (1992), in their analysis of transport geography at its intellectual height, note that transport geographers have therefore coalesced around the following themes:

- Transport and development: colonial expansion and (un)even development;
- Transport policy and control;
- Transport, environment and energy;
- Urban transport patterns and problems;
- Rural transport;
- Transport, leisure and tourism;
- Freight transport;
- International transport for passengers and freight.

This list of themes perhaps reveals where the key academic trajectories of conventional transport geography have been focused; that is on up-scaled and macro-level system and network analysis, and with the traditional interests of geographers in terms of analysing particular types of space, such as the urban and the rural. In this way, conventional transport geography has, on the most part, taken its intellectual inspiration from the quantitative revolution in geography and the resultant focus on positivist approaches, which upholds the scientific method as a mode of inquiry (Shaw and Hesse, 2010). In inhabiting this intellectual space, transport geography has drawn on other positivistic social science disciplines and to some extent from the engineering and physical sciences to develop concepts and methods for analysis. In large part, these have followed a quantitative methodological tradition. Examples from the early editions of the *Journal of Transport Geography* attest to this kind of approach and include:

- Ivy's (1993) connectivity analysis of US domestic air transport networks, which utilised mathematical approaches to derive a classification scheme for air transport connectivity;
- Luoma *et al.*'s (1993) deployment of a mathematical threshold gravity model to examine the role of distance in determining transport needs as a planning tool;
- Bassett's (1993) description and analysis of British port privatisation during the 1980s and the resultant economic impacts;
- Dundon-Smith and Gibb's (1994) analysis of the economic impacts of the Channel Tunnel crossing through economic potential analysis;
- Hine and Russell's (1993) study of pedestrian crossing behaviour in Edinburgh, which used quantitative measures of traffic flows, crossing behaviour and visitor perceptions of risk using questionnaire surveys.

Transport geography has therefore been concerned with the adoption of traditional scientific methods and principles to explore and analyse both transport networks and the impact of these networks on economic activity and human populations. The focus lies particularly in satisfying the requirements of traditional scientific approaches, namely objectivity (analysing without subjective judgement), validity (ensuring measures are accurate and valid), hypothesis testing (using analytical techniques to try to over-turn an idea using quantitative evidence) and reliability (ensuring that the research is replicable by others and should achieve similar results if undertaken again). Indeed, as with other kinds of positivist research, such studies have sought to be representative of the populations being studied, hence the adoption of quantitative methods and the use of particular types of statistical analysis appropriate to probability-based data.

Case study: rural transport accessibility

As a way of illustrating and exploring the ways in which transport geographers have examined key themes, we can use a case study approach to highlight the

essential elements of a conventional transport geography approach for understanding the relationship between space and accessibility. As a guide, Nutley's
(1992) overview of what he terms the rural 'accessibility problem' will be
explored through illustrating some of the key historical concerns of rural transport provision and also the links transport geography has had to the emergent sub-
discipline of rural geography. Of its time, this set of research clearly illustrates the
ways in which transport geographers have explored the issue of rural transport
and what follows is very much a time-limited analysis from the 1980s and early
1990s. As such, it is purely illustrative of a transport geography approach and we
want to emphasise that the broader intellectual trajectory of transport geography
still adopts the basic epistemological assumptions laid out below, albeit with reference to other issues, such as research on consumer travel preferences (Hensher
et al., 2015), logistics analysis (Sakai *et al.*, 2015) and road network analysis
(Freiria *et al.*, 2015).

Rural transport has been regarded for many years as a key 'problem' (Cloke,
1993) and one that has been highlighted recently in the UK with cuts to local
authority spending and the resultant need to reduce services (*The Guardian,* 2015).
As Cloke (1993) has noted, much of the discourse underpinning this problem has
been the connectivity of rural areas to urban centres and the economic benefits this
can accrue. And it is this underpinning policy logic that sets the context for a discussion of the ways in which rural transport has been regarded in policy settings
since the Second World War.

The 'Golden Age', as it is seen, for rural accessibility is often regarded as the
period between the two World Wars, when a plethora of rural rail lines criss-
crossed the countryside and brought relatively cheap connectivity to the UK's
train network. A look at the Great Western Railway map of Devon from the 1920s
shows the full extent of branch lines to rural locations, even reaching the heights
of Princetown on the top of Dartmoor and connecting small hamlets and villages through which only unclassified roads now run (Figure 4.1). This period
of railway expansion and connectivity was supported by a host of onward travel
options and services, supplied by both horse-drawn power and road vehicles and
coaches.

However, as Nutley (1992) notes, car ownership began to expand dramatically
in the 1950s. As we noted in Chapter 2, the rise and popularity of the motor car
as the preferred mode of transport was not simply an economic choice (though
this was clearly important) but was part of a liberalisation process that matched
the growth in the economy in the UK and a rising demand for mobility as technological developments began to change the structure of British employment. The
effect of this on the countryside was dramatic and confirmed a long-term trend
that saw a shift in the UK workforce dominated by agriculture (21 per cent of
the workforce in 1841) and manufacturing (36 per cent of the workforce in 1841)
towards services (81 per cent in 2011, with just 1 per cent being in agriculture)
(Office for National Statistics, 2013). Yet despite the radical de-population of rural
areas by former agriculture workers, there has not been a collapse in rural society,
with agricultural labour being in part replaced by intensive and technology-reliant

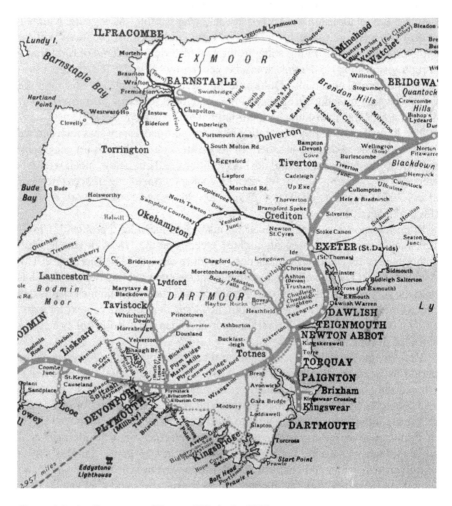

Figure 4.1. A railway map of Devon, UK, in the 1920s.

Source: Great Western Railway. Copy of a Great Western Railway enamel sign. Public domain, from Wikimedia Commons.

farming practices, the development of rural service economies, most notably tourism and leisure, and the gradual spread of home-working as telecommunications improve (Gilg, 1996).

From a transport perspective, what these changes have meant is a radical modal shift from railway transport to that of the car and lorry. Perhaps the most infamous point along this journey was the highly controversial closure of railways in the early 1960s as the result of Dr Richard Beeching's report, *The Reshaping of British Railways* (British Railways Board, 1963). In this report, Beeching identified over 2,000 stations and 5,000 miles of railway for closure, in large part to

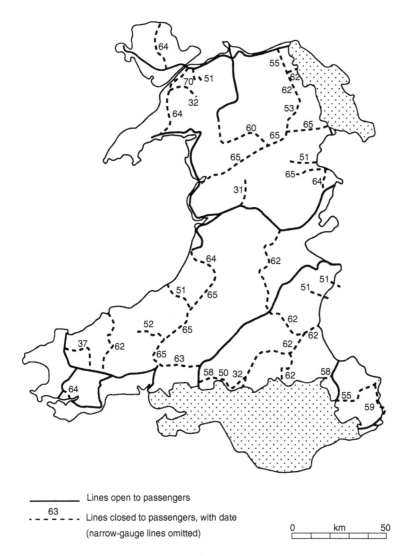

Figure 4.2. Railway closures in Wales (urban north-east and south excluded) (based on Nutley, 1982).

reduce the losses being experienced by the nationalised British Railways, which was competing hard for market share against road transport.

Nutley (1992) highlights the radical change in rural rail transport that was brought about by line closures, as evidenced in Figure 4.2. It is interesting to note that many of the lines included those that were unviable even before Dr Beeching's report. Similarly in Devon, the only remaining lines today are the mainline through the county and branches to Barnstaple, Exmouth and Gunnislake. As a result, rural transportation has largely become an issue of road transport and the accessibility issues surrounding car ownership on the one hand and bus services on the other.

Within the context of rural accessibility studies, transport geographers have adopted a number of approaches to study and critically appraise the relationships between car ownership, access and economic development in geographical context. First, Nutley (1985, 1992) has explored the idea of accessibility within the context of need and deployed the use of a time–space approach to examine how accessible services are to residents living in villages within a specific geographical area. As Nutley (1985) noted in developing this methodology, aspects of social need are normative in nature and are highly objectified, and include travel for work, shopping, leisure, healthcare and so on. Within village populations, categories were then formed to represent different groups: elderly, working people, housewives (sic) and children. These were then further divided into car and non-car owning groups. The resultant analyses for villages in Wales provided a set of accessibility curves, an example of which is shown in Figure 4.3. Not surprisingly, those without cars had the greatest challenge to access services, but it was clearly young people who were most disadvantaged, with limited access to a wide range of services. Such an approach, based on the modelling of objectified needs-based criteria can have a role to play in farming planning choices for providing rural service needs, although as Nutley (1985) highlighted:

> The method makes assumptions and has a high degree of subjectivity, and this means that its usefulness stands or falls on the consistency with which it is applied. Whatever value judgements are implicit in the initial parameters, consistent application of techniques of this sort will provide a clear statement of the relative advantages of alternative policies, and of the consequences for the people affected.
>
> (1985, p. 49)

Second, Nutley (1992) has outlined a range of ways in which researchers can explore rural transport issues through discrete intellectual and methodological lenses:

- Regional scale analyses of general accessibility to transport that draw on simplified data to provide indices of transport 'deprivation' in particular areas, which can be analysed spatially and according to administrative boundaries;
- Analyses of individual aspects of personal accessibility through exploring the social dimensions of access and the ways in which particular groups can become marginalised and excluded from transport accessibility;
- Historical analyses of transport accessibility trends in the context of particular places and identifying the reasons for these shifts;
- Policy analyses and planning scenarios that use 'before and after' studies of the impact of particular interventions in local areas and how new policies might affect accessibility.

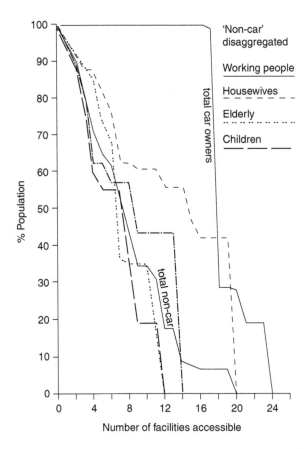

Figure 4.3. Local accessibility by social group, Radnor District, Powys, Wales, 1981 (based on Nutley, 1985).

Accordingly, in Nutley's (1992) analysis of the ways in which transport geographers can deal with issues of rural transport, accessibility and mobility, the emphasis was very much on understanding spatial trends, relating this to social inequalities in accessibility and appreciating the macro-scale impact of policies to effect change. As we shall go on to see later in this chapter, these types of studies have been based on a series of epistemological assumptions that have come into question as researchers apply different theoretical frameworks and methods to address the issues mentioned above. However, what is clear is that conventional approaches to transport geography have very much focused on the application of quantitative and scientifically replicable approaches to studying transport. In the next section, we explore how the 1990s heralded a change in the ways in which transport and mobility were viewed in the wider context of change in the social sciences.

Transport geography and the 'mobilities turn'

The dominance of quantitative transport geography until the 1990s is to some extent representative of the broader academic landscape of geography up to this time. However, geography's shift towards what has been termed a 'cultural turn' (Cloke *et al.*, 2007) was witnessed in transport geography not by an over-turning of existing practice, but rather by an importing of new ideas and methods from an interest in disciplines like sociology in mobility. As such, the stage was set for the emergence of two quite distinct ways of practicing an interest in transport and mobility by geographers (Shaw and Hesse, 2010), which has carved out the landscape that we see today.

Knowles *et al.* (2008) note the major change in emphasis in the way transport geographers conceived of their sub-discipline was to do in part with external economic conditions, but in large part because of a changing intellectual landscape in the social sciences:

> … various factors, including the advent of cheap oil and a reluctance on the part of transport geographers to engage in significant theoretical debates, led to too many human geographers significantly downplaying transport matters in their analyses of social and economic patterns and systems.
>
> (Knowles *et al.*, 2008, p. 3)

Indeed, in Keeling's (2007) review of progress in transport geography research, he was compelled to note that certain aspects of the sub-discipline's progress had been less than inspiring, highlighting the mundane and static nature of certain research fields. Moreover, Hall *et al.*'s (2006) analysis of why transport geography had become so irrelevant to new forms of economic geography was equally critical and was based on the views of those brought together at the Annual Meeting of the Association of American Geography in 2004:

> Transport geography, it was noted by some on the panel, can be characterized by a strong empirical tradition, a mostly positivist approach, and a certain lack of theoretical orientation. In contrast, much economic geography is more theoretical, abstract, and, in keeping with the wider trends in other subdisciplines of human geography, some economic geographers have already opted for a positivistic, 'interpretative' orientation of the field. This is by no means the case for transport geography.
>
> (Hall *et al.*, 2006, p. 1402)

The essence, therefore, of arguments emerging in transport geography was about the prominence of transport geography as part of the wider discipline of human geography and how transport geographers could deal with a whole host of 'new' issues being raised by scholars from others disciplines, and being used by human geographers to reframe debates on mobility. This essence of the

intellectual debate was framed by a ground-breaking paper on the new mobilities paradigm, published in 2006 by Sheller and Urry, in which they outlined
the alternative narrative to conventional transport geography. The key argument
they made was that social science, for which we can also read conventional
transport geography, had become static and incapable of dealing with the social
complexity of mobility:

> Travel has been for the social sciences seen as a black box, a neutral set of tech-
> nologies and processes predominantly permitting forms of economic, social,
> and political life that are seen as explicable in terms of other, more causally
> powerful processes. As we shall argue, however, accounting for mobilities
> in the fullest sense challenges social science to change both the objects of its
> inquiries and the methodologies for research.
>
> (Sheller and Urry, 2006, p. 208)

In adopting this position, Sheller and Urry (2006) argued that sedentrism was to
suppose that nearly everything that gives meaning to life and for which we should
seek to provide an analysis rests on specific places. Rather, they argued, what is of
interest is how movement, mobility and travel are of importance and have meaning, and act to (re)produce social relations. What we had forgotten, so the argument went, was the meaning of movement both in a physical and a virtual sense.

As we shall see in Chapter 8, the intellectual landscape of transport and
mobility is one that now reflects the broader topography of human geography
and social science more generally, with at least a significant divide between posi-
tivist/quantitative approaches and those which adopt post-positivist/interpretivist
and qualitative perspectives. Using the car and auto-mobility as an illustrative
example, Laurier (2011), Merriman (2009), Sheller (2004) and Sheller and Urry
(2006) have all argued that previous conceptualisations of car transport and travel
have tended to emphasise an objectified approach to exploring either the impact
of car transport in comparison to other modes or the rationalistic ways in which
people make decisions about car use, journey length and purpose, and the alternatives available, perhaps even taking into account the environmental consequences.
However, all of these approaches assume that the journey itself and the movement
embodied in travelling have no interest or consequence. In other words, when we
think of the car, car travel and what it involves, we can find renewed meaning for
both individuals and the wider social consequences of auto-mobility. This might
be about how identities are framed and reinforced through car use, the social networks and bonds that are formed through travelling with others, the ways in which
spaces are apparently liberated through auto-mobility, and the social practices that
develop to reflect and reproduce economic patterns of organisations, such as work
commuting and sub-urbanisation. As such, advocates of the new mobilities paradigm have urged scholars to focus on the social dimensions of transport and to do
so through interpretative methods that enable an uncovering of the dynamism and
complexity of movement.

The new mobilities paradigm and geographies of mobilities

At its heart, most of the interest in the new mobilities paradigm has started from the perspective that life is, as Adey (2010) put it, moving and is probably moving more than it did before. Urry (2011) went further and addressed the notion of hyper-mobility in what he terms a 'speeded up' world. The examples he quoted are adequately illustrative of his point about 'fast travel':

- The growth of low-cost airlines to provide regular and cheap travel;
- The growth of auto-mobility in countries with large populations, such as China and India;
- The development of high-speed rail in Europe;
- The globalisation of tourist attractions that require long-distance travel;
- The rise in long commuting distances for work;
- The increases in global carbon emissions associated with these practices.

We might add to this list, as we shall explore further later on, the rapid growth in virtual movement and communication, which is revolutionising social inter-action and presenting us with the potential to be perpetually connected and instantly available, a change in the past 15 years that is probably not unlike the revolution experienced by those who first experienced the opportunity to leave their home town to travel by the railway at speeds unimaginable just years before.

To characterise the new mobilities paradigm we need to recognise that mobility is not a word solely associated with transport, in the traditional sense. Indeed, Urry (2007) demonstrated that at its most basic, mobility can be interpreted in several different ways:

- Movement and hyper-mobility: the movement of subjects through physical or virtual space and, in the context of hyper-mobility, the sense that we are becoming ever more mobile;
- A 'mob' that is constantly changing: the changing nature of movement in a social context. For example, in moving through physical and virtual spaces, we frequently change our group associations and networks;
- Social mobility: the ways in which individuals move through social groups, traditionally expressed as 'up' or 'down' social hierarchies or classes;
- Semi-permanent geographical movement: the trend for more people to relo-cate for work or other purposes on a semi-permanent basis.

As such, mobility is a broad and complex term and in the context of trans-port we would tend to focus more on the first category as a mode of analysis, while recognising that this does have impacts for the other three types of mobility. Within this context, Urry (2011) has argued that mobility needs to be thought of as essentially corporeal, that is movement which is embodied and the analysis of which enables us to appreciate the deeper meanings of mobility for the individuals concerned. Such corporeality sits alongside the physical movement of objects, services and information as other discrete forms of mobility.

The mobilities paradigm responds to these definitions and classifications of mobility by providing an organising framework for research, although there is no intention to provide a single theory of what Sheller and Urry (2006) termed a 'grand narrative'. Indeed, as Adey (2010) has highlighted, the mobilities paradigm is not about occupying theoretical ground and colonising geographical themes, but rather it is about highlighting where mobility matters and upholding the study of movement as geographically significant.

As an organising framework, Hannam *et al.* (2006) present an overview of how we can approach the study of mobility. First, they argue that attending to the linked themes of migration, travel and tourism enables researchers to explore both the social value of travel and also the underpinning political geographies that promote or even compel travel, and the governing frameworks in which such mobility operates:

> This implies attending to obligatory as well as voluntary forms of travel. In many cases travel is necessary for social life, enabling complex connections to be made, often as a matter of social, or political, obligation.
>
> (Hannam *et al.*, 2006, p. 10)

Second, virtual and informational mobilities are critical for our contemporary understanding of how lives are being shaped by the new technologies of instant communication and visualisation, and the role that immobile infrastructure, such as mobile phone masts plays. Anecdotally it is clear that relationships within social networks and the practices that govern such interactions are rapidly evolving as mobile technologies foster new ways of engaging in both social and political actions.

Third, Hannam *et al.* (2006) highlight the importance of key spaces of interaction during the process of mobility. Such nodes in a network provide opportunities for sociability and include airports, railway stations, bus stops, cafes and metro stations, alongside places of leisure and tourism, such as beaches, attractions, theme parks and so on. Indeed, in analysing mobility nodes and spatial mobilities, the point is made that travel time has much value and that such time can afford an engagement with practices and interactions that are unavailable in other contexts.

Finally, materialities need to be considered as key attributes of understanding mobility, from the ways in which new technologies are developed and adopted, to appreciating the hybrid nature of the materials involved in mobility. As Hannam *et al.* (2006) note 'Crucial to the recognition of the materialities of mobilities is the recentring of the corporeal body as an affective vehicle through which we sense place and movement, and construct emotional geographies' (2006, p. 14).

Accordingly, mobility is about uncovering the complex social processes that govern the ways we move and the consequences of that movement for understanding how we might conceive of future mobilities. Conceptually, Cresswell (2010, 2011) and Cresswell and Merriman (2011) have been formative in interpreting these calls for geographers to recognise the value of mobilities research into a broad theoretical structure. They have argued for a focus on three key aspects, which have formed much of the basis for geographical scholarship on mobilities in recent years. First, practices highlight the ways in which we have developed forms of movement in our everyday lives: walking, cycling, running, driving, flying and so on. In the context of this book, practices are perhaps the most fundamental building block of mobility and its most easily recognisable form. Second, Hannam *et al.* (2006) have highlighted the role of 'moorings' in the context of mobility; that is, the need to consider mobilities alongside the infrastructural spaces that service mobility: ports, airports, stations, car parks and so on. Cresswell and Merriman (2011) refer to these, in geographical context, as 'spaces', which 'produce structural or infrastructural contexts for the practising of mobility' (p. 70). Third, 'subjects' are critical to an understanding of mobilities; we as 'citizens' practice mobility, but we need to appreciate the ways in which we adopt particular roles in different mobility contexts: commuters, school runners, leisure cyclists, back packers. Indeed, as we shall go on to see in Chapter 8, we also need to appreciate the ways in which our practices can represent different characteristics of our identities in alternative sites of practice; we are a low-carbon 'environmentalist' for our everyday travel and a carbon-intensive 'polluter' when on holiday (Barr *et al.*, 2010).

Using Cresswell and Merriman's (2011) classification, we can begin to see the rich body of work that has emerged in the past 15 years in geography and related disciplines such as sociology and anthropology, which has all contributed to our understanding of practices, spaces and subjects.

Beginning with practices, this is perhaps the field that has attracted greatest attention, in part because it has challenged traditional notions of behaviour and behavioural change, for so long the mainstay of psychological understandings of individual actions (Owens, 2000; Shove, 2003, 2010). As we will describe in more detail in Chapter 8, the dominant epistemological tradition in social science for analysing how people behave and why they do has been driven by a social psychological perspective, which has adopted a cognitive and individualistic approach towards understanding behaviour. In this way, behavioural research has been and is still dominated by the use of theoretical frameworks of individual decision-making (e.g. Ajzen, 1991; Fishbein and Ajzen, 1975). As we go on to evidence in Chapters 5 and 6, this tradition is still popular in certain parts of transport geography (e.g. Heath and Gifford, 2002; Kenyon and Lyons, 2003). However, sociologists in particular have challenged the basic assumptions underpinning these studies, which have been cast as focusing too greatly on individualistic, cognitive and rationalistic decision-making processes, beyond the broader social context of what drives behaviour.

Accordingly, sociologists, and now geographers, have adopted an alternative perspective on behaviour that has utilised Reckwitz's (2002) overview of practices (Warde, 2014). Although practice theory is complex and there are indeed a range of perspectives on what has become known as social practice theory (Shove *et al.*, 2012), the essence of studying practices is about seeing what appears to be individual behaviour (as a result of cognitive and rationalistic decision-making) as part of a wider set of social contexts and narratives about how to behave, hence the use of 'practice' as a linking term between individualistic elements and the role of social influence and underpinning economic conditions.

A useful example of how we might consider the difference between the social-psychological and sociological perspectives in the behaviour versus practice debate is the use of motor cars for getting to and from work. Figure 4.4 presents a typical image of a busy freeway in the USA at a peak during the working day. From a behavioural perspective, we can see this scene as so many transport geographers have done as an example of hundreds of individual decisions to take the car to work on a given day of the week. This might, from this perspective, be influenced by a whole range of factors, from weather conditions, fuel price relative to public transport use, perceptions of convenience and comfort, all the way through to personal considerations about carbon footprints and air pollution. As we shall see in Chapter 8, seen from this perspective, trying to change driving behaviour becomes an issue of attempting to engage directly with the individual driver and to change their perceptions through effective information and communication.

Figure 4.4. A typical view of a freeway in the USA on a work day.

Source: Minesweeper on en.wikipedia.

By contrast, a practices approach does not view the scene in Figure 4.4 as one depicting individual decisions alone, but rather views the collection of vehicles as representative of the social practice of commuting. Accordingly, social practice theorists would argue that driving to work – the act of commuting – is constitutive of a broad socio-economic phenomenon that has developed over the past century, in which it has become both commonplace and desirable to live further from work, as aspirations to reside in suburbs has become popular and the motor car – and its attendant infrastructure – has enabled cheap and relatively rapid commuting. Therefore, commuting to work by car has emerged through time as a practice and as a consequence cannot be shifted easily by simply providing more information or better communication to the individual; rather, it is about altering the contexts in which people live and work.

Research focusing on practices in mobility studies has been extensive and has covered a wide range of travel modes. Indeed, what is intriguing and fascinating about research from the new mobilities paradigm is the valuable upholding of geographies of the everyday – an appreciation of the practices we take for granted (Holloway and Hubbard, 2001), such as walking (Lorimer, 2011; Lorimer and Lund, 2003; Middleton, 2009, 2010; Weilenmann *et al.*, 2013), cycling (Spinney, 2008, 2009, 2010, 2011), driving (Laurier, 2011; Laurier *et al.*, 2008) and flying (Budd and Adey, 2009; De Lyser, 2011). Indeed, collections such as that edited by Vannini (2009) have highlighted the cultural richness of mobility through particular practices, such as making railway journeys (Bissell, 2009a, 2009b) and something as apparently mundane as a bus journey (Jain, 2009).

Accordingly, the study of mobility practices has undoubtedly provided a rich context for sociologists and geographers to appreciate both everyday and tourism mobility in its rich complexity. The study of spaces, particularly by geographers, has also been a rich ground for appreciating how mobility (re)produces space and the ways in which mobility spaces are constructed, engineered and managed to become affective in nature (Adey, 2008, 2010). Researchers exploring spaces have tended to focus on the infrastructures that enable or inhibit personal mobility: airports (Adey, 2008), roads (Merriman, 2009) and metro stations (Dunckel Graglia, 2016). Indeed, the focus of such research has been to unveil the meanings associated with apparently transitory spaces and to explore how these can be infused with power and control.

A useful example is Adey's (2008) research on the ways in which airports are designed and engineered to produce particular affects in the ways that people move and respond in an airport setting. He advocates a far closer link between geographies of mobility and geographies of affect to 'develop an understanding of architecture as a situational affective context that lays down root textures and motivations for movement and feelings' (2008, p. 439).

In doing so, Adey explores how airports are seeking to promote what he terms 'predictive passengers' through the ways in which architectures and practices are mediated through the demands of the modern airport. For example, the dual objectives of security and commercialisation have meant that 'Airports and other borders work to construct and channel the possibilities of the sensations and emotions experienced within these sites' (2008, p. 442).

Airports thus attempt to use architectural regimes to suppress certain emotions and sensations, while accentuating others in the dual aims of reassurance and promoting passenger spend. In this way, airports have invested considerable amounts of money to understand the characteristics of different types of passenger in order to predict their likely behaviour and movements. All of this leads us to appreciate the connections between mobility, architecture and affect and the ways in which spaces of mobility can be controlled and placed into regimes that govern movement.

Finally, a focus on subjects requires us to concentrate on who the people are who are moving. Researchers adopting the new mobilities paradigm have made great efforts to contextualise discrete and objectified categorisations of travellers – workers, holidaymakers, school runners, those visiting friends and relatives – in order to appreciate how mobile identities are formed and how these relate to practices. Accordingly, a focus on subjects is about moving beyond the individual and appreciating how representations are formed and perpetuated. Examples of such research have included both subjects in the everyday, such as Edensor's research on commuting (2011, 2012), alongside those who represent Urry's (2007) semi-permanent mobile subject, the refugee (Ashutosh and Mountz, 2012). In these accounts, emphasis is placed on how the act of moving is constitutive of a particular identity and how this influences practice. Indeed, in accounts of relating subjects to broader notions of lifestyles (Vannini, 2009), there has also been a focus on community and the shared practices that emerge in particular groups, such as Kleinert's (2009) analysis of those in the cruising community and Rickly-Boyd's (2013) exploration of rock climbing as performance of identity.

One example of the ways in which subjects have been explored is through examining some of the apparent contradictions that can emerge from adopting typologies, a theme which we shall explore in further detail in Chapter 8. Barr and Prillwitz (2014) have built on research in tourism studies that has indicated a high degree of scepticism about the role of tourism travel in producing harmful carbon emissions that contribute to climate change (Gössling and Peeters, 2007) and in so doing have highlighted how different sites lead to radically differing practices (Barr *et al.*, 2011b). Accordingly, Barr and Prillwitz (2014) argue that tourists have to negotiate a different identify frame between everyday and touristic practices. While in the everyday context, subjects were able to match their aspirations to be regarded as 'good' citizens by engaging in various pro-environmental practices, such as cycling, walking, energy saving and water conservation. Yet these 'good' behaviours did not transfer into touristic sites of practice, where the desirability to engage in conspicuous forms of consumption and to use these for identity reinforcement came into conflict with pro-environmental concerns. As such, tourists were able to rationalise carbon intensive activities like flying by either arguing that these were offset by other practices, or were simply too important to forego. In this way, subjects were able to hold alternative identities in the context of climate change and practice according to the consumption context:

> ... climate change poses a significant challenge for both researchers and policy-makers as they attempt to understand the ways in which 'citizen-consumers' can become agents for and of change [amid] increasing evidence showing that the challenge of climate change is seemingly irreconcilable with current demands for mobility in the 21st Century.
>
> (Barr and Prillwitz, 2014, p. 234)

The emphasis in the new mobilities paradigm on practices, spaces and subjects therefore affords researchers considerable scope for understanding the changing nature of mobility in the twenty-first century, and to appreciate how moving has meaning and relates to the structures and infrastructures of a mobile society.

Conclusion: transport geography and geographies of mobilities?

It is clear from the preceding two sections that conventional approaches towards studying transport geography and the emergence of the new mobilities paradigm in the social sciences represents a challenge for contemporary transport geography, as an identity-framing exercise for the sub-discipline. The differences do appear stark; conventional approaches to transport geography adopt a largely positivist epistemology, which invokes scientific principles and aims to use rationalistic reasoning at all scales of analysis. Even when studies of travel behaviour are considered, there is a tendency to adopt logical positivist frameworks of analysis to uncover the factors that influence human behaviour. In contrast, the new mobilities paradigm invokes an interpretivist and post-positivist paradigm that seeks to uphold and work with subjectivities and to question the use of representativeness as a basic principle. Indeed, these epistemological concerns are reflected in the methods utilised by the different traditions. Conventional transport geography has largely remained rooted in the quantitative tradition, while researchers using the new mobilities paradigm have deployed qualitative methods, such as ethnography, visual and textual analysis. Moreover, it could be argued that the two approaches are often seeking to address quite different questions.

Within transport geography the concern often lies with understanding how networks develop, the impact of these networks, their relationship with economic development, and the articulation of these findings for policy makers. Mobilities research has often shied away from trying to develop narrative for policy making, perhaps because the dominant tradition in central and local government is driven from economic analyses. As such, mobilities researchers have concentrated on arguing for much richer and often complex (and certainly politically problematic) understandings of mobility. As we shall see in Chapters 8 and 9, depending on the perspective adopted, approaches for promoting sustainable mobility vary radically and offer very different outcomes.

It is, however, not the case that the divide between these two different approaches has been ignored by those in transport geography with a concern for promoting the sub-discipline to the wider geographical community. Most notably, Shaw and Hesse (2010) and Shaw and Sidaway (2010) have attempted to explore the apparently yawning gap between transport geography and mobilities research by arguing that 'while these two different strands of thought and scholarship have been practised more-or-less separately, there ought to be closer working – or at least a better understanding – between mobilities and transport geographers' (Shaw and Hesse, 2010, p. 305).

The *modus operandi* that Shaw and Hesse (2010) advocate for appreciating the commonalities between these two approaches was based on a panel convened at the Annual Meeting of the Association of American Geographers in 2008. First, they argue that the new mobilities paradigm has brought new energy and insight into debates in transport geography by focusing on the experiential components of travel and exploring how meaning is brought to transport through mobility. Indeed, at the heart of the commonalities between transport geography and mobilities research is the theme of movement, which clearly unites both areas of research. Within this context, Shaw and Hesse (2010) advocate the following themes to pursue mutually beneficial research agendas:

- Mobility should become a more central component of transport research and for geography as a whole, given the importance of understanding the process of moving between places across space;
- Transport geographers and mobilities researchers should make greater efforts to explore the common theoretical and methodological approaches they can take to deal with issues like the understanding of daily mobility, which inevitably involves both understandings of individual behaviour and broader understandings of social practice;
- Policy and the importance of transport and mobility for creating sustainable futures should be a point of coalescence and one avenue to pursue would be broader socio-economic analyses of policies that drive and influence particular kinds of practice.

These calls for mutual recognition and a closer working are certainly well-founded, because it does the sub-discipline of transport geography no good at all to provide a divided and apparently irreconcilable set of assumptions and working practices to both the wider geographical community and to publics and policy-makers. Indeed, it is likely that movements within countries such as the UK to place a greater focus on research impact and collaborative working between academic researchers and policy-makers and practitioners could lead to common practices that speak to broader agendas. However, it is also the case that the epistemological differences are so fundamental that it is likely there will be different approaches to the basic challenges of transport and mobility for some time yet.

We might therefore ask: where does this take us to? In this chapter we have sought to provide an overview of the fundamental approaches underpinning the practice of transport geography and mobilities research. Our aim has very much been to present these intellectual landscapes as they exist, rather than to make a snap judgement on either. Instead, we have used this exploration of transport geography and mobilities as a way of foregrounding two contexts that we now move onto: the everyday and the touristic. In Chapter 5 we examine the ways in which researchers have sought to understand the complexity of daily travel, while in Chapter 6 we explore the nature of tourism travel as something that has gained huge prominence during the twentieth and early twenty-first century.

5 Travel and transport in everyday life

Introduction

This chapter is the first of two that explore the spatial contexts of transport and mobility. We have specifically chosen the everyday (this chapter) and tourism and leisure contexts (Chapter 6) because they afford spatial lenses through which to view individual travel in the twenty-first century. While tourism and leisure travel has a specific purpose, the theme of everyday travel relates to a range of regular practices that are often habitual in nature and which connect to routine economic, social and cultural activities. These are commonly attributed to categories such as daily commuting, the 'school run', shopping trips and visiting friends and relatives. Yet these categories can hardly capture the complexity of what are a coalescence of pre-formed and in-situ sets of 'decisions', all of which make understanding travel behaviour highly complex and hotly contested.

As we noted in Chapters 2 and 3, everyday travel is something that has changed radically, even within generations. Even if we exclude the ways in which people move in the virtual world with the advent of mobile technologies and the Internet, the ways in which we have commuted to work, school and used retail outlets have all radically shifted. Take for example the shift in work commuting for many people in the UK. For many people in manual and semi-skilled roles, both in urban and rural areas, travel to work would have been a relatively short commute or would involve no travel at all. During the period of industrialisation and even until the 1980s it was common for factories, coal mines, steel plants and so on to have dedicated housing for workers in close proximity. Indeed, this meant that most workers travelled to and from work on foot or by bicycle, and did so at the same time of day. Even holidays were predetermined by a factory close down. These kinds of centralised and up-scaled modes of economic production led to mass mobility, but within narrow spatial confines, with a common mode of transport and shared practices and norms. Contrast this with the situation in the UK today, where there are relatively few large industrial plants, but rather a diverse and spatially distributed set of service-based organisations, often located on cheaper land on the outskirts of settlements, where developers have planned in large car parks, and where it is only with some persuasion that public transport is provided for the few who deign to break the social norm. In this way, going to work has become an exercise in personalised travel planning and an individual-ised experience, determined in large part by personal calculations of priorities: family commitments, picking up children from day care or school, purchasing goods from the supermarket and negotiating 'core hours' with employers. In short, our daily travel lives are now highly personalised and represent a broader shift on economic practices from shared modes of production to highly specialised and bespoke economic activities.

The study of everyday transport and mobility is therefore the study of people and their interaction with space, technologies and infrastructures. Most impor-tantly, it is an analysis of how broader economic and social trends have influenced these spaces, technologies and infrastructures, which lead to rapidly evolving travel practices. In this chapter, we explore the ways in which geographers have examined everyday transport and mobility through the two epistemological and methodological lenses we outlined in Chapter 4. In the first instance, we trace the heritage of psychological research in travel behaviour studies, which has resulted in a wealth of knowledge concerning the motivations and barriers for adopting particular behaviours. In particular, we explore some of the theoretical models that have been applied by social psychologists to understand travel behaviour and the ways in which travel behaviour research has been conducted through quan-titative survey instruments that rely on a psychometric and cognitive approach.

In exploring the second approach towards examining travel practices, we deploy theoretical and methodological ideas from the sociological realm to exam-ine how mobilities researchers have understood travel 'decision-making'. In so doing, we explore the ways in which social practice theory can be used to appreci-ate the contextual and situated nature of mobility practices, which can be viewed not as the result of individualised and preformed decision making, but rather as representations of wider social practices. Thus we examine common notions such as the apparent 'decision' to travel to work each day using a car, versus a refram-ing of this behaviour around the notion of 'commuting' as a shared and therefore social practice, which is shaped by broader economic and social contexts.

The chapter ends by highlighting the challenge of understanding everyday travel from the perspective of each epistemological position and we argue that there is a growing tension between academic debates on travel practices and the needs of practitioners, who are eager to understand everyday travel and, poten-tially, to try and change it, a theme which we develop in Chapter 7.

Travel behaviour: values, attitudes and actions

Attempts to understand what we can broadly term everyday travel behaviour have been ongoing since the 1970s (e.g. Heggie, 1978) and although the focus of this particular book lies in appreciating the interaction between individuals, society and the emergent issue of anthropogenic climate change, it is only recently that researchers have sought to deploy the theories and methods applied in travel behaviour research to respond directly to such global environmental concerns. Accordingly, the long heritage of travel behaviour research has been concerned with far wider and more basic questions about how to define travel behaviour, what factors appear to influence travel decision-making and how frameworks can be developed to promote behavioural change (an issue we shall return to in Chapter 8). However, the voluminous collections of research accumulated over decades of travel do reflect one, and only one, intellectual position that has come to dominate, until recent years, the landscape of research into the social dimensions of transport and mobility. It is to this set of intellectual assumptions that we initially turn.

As noted in Chapter 4, transport geography in general has been marked out until recent years by a set of conventional epistemological assumptions that have largely drawn on a scientific tradition (Knowles *et al.*, 2008; Shaw and Hesse, 2010). This positivist and largely quantitative approach to research is very much reflected in the intellectual trajectory that has been adopted by those researchers exploring the specific issue of travel behaviour. We can characterise this approach towards travel behaviour in the following way by drawing on research from the wider behavioural sciences (e.g. Slovic, 2000), commentaries from the environmental social sciences (e.g. Burgess *et al.*, 1998; Owens, 2000), alongside broader research texts in the social sciences (e.g. Bryman, 2011; Kitchin and Tate, 2013).

First, the disciplinary background to travel behaviour research has largely been from psychology, in particular elements of psychology that have drawn on cognitive approaches to understanding behaviour and which have adopted psychometric approaches to measuring attitudes and perceptions (Anable, 2005). Adopting a psychological perspective has meant that researchers have largely drawn on particular social-psychological theoretical frameworks to undertake their research.

Second, this focus on a theory-led approach has naturally utilised an inductive mode of research inquiry, where theoretical narratives are used to construct research designs, survey instruments and experiments. In accordance with positivist approaches, the aim then becomes to construct hypotheses to accept or reject using the data collected.

Third, therefore, the types of data collected are required to be quantitative in nature, in large part to satisfy the scientific basis of the research being undertaken, which implies that samples of individuals being studied are selected by probabilistic means to be representative of the wider population. Indeed, there are also requirements that the study instruments used are 'valid' (measuring constructs they are intended to measure) and 'reliable' (they will be repeatable between studies).

Fourth, in undertaking research of this kind, positivist approaches assume a level of objective truth that travel behaviour researchers have attempted to maintain in their work. Indeed, we can extend the notion of objectivity in research to the broadly rationalistic paradigm that has dominated such scholarship, through a reliance on assuming that decisions are based on linear forms of reasoning and that it is indeed possible to identify, measure, analyse and come to a definitive view on the factors that influence travel behaviour decision-making.

Finally, transport behaviour research has tended to be avowedly individualistic in nature; that is, has tended to focus on the cognitive aspects of the ways in which individuals receive information and make choices. Although external factors can be accounted for in rationalising such choices, the emphasis is very much on viewing behaviours as choices made by individuals, rather than regarding behaviours as constitutive of broader social practices that are historically rooted and shared.

Accordingly, travel behaviour research, like so much other scholarship in the behavioural sciences (see reviews by Lupton, 2013; Slovic, 2000), has been concerned with attempting to provide quantitatively reliable and verifiable data on behaviour and the attributes that appear to influence actions. In this way, the hope has been, it might be possible to appreciate how to manipulate particular attributes, such as negative attitudes or perception, in order to overcome what are often termed 'barriers' to behavioural change. It is with this intellectual lens that we now take a look at the ways in which transport geographers have framed travel behaviour, before exploring some examples of how this has been deployed empirically.

Defining travel behaviour

For transport researchers, the basis for defining travel behaviour is grounded in an understanding of need. A fundamental condition for human existence is the satisfaction of individual needs, ranging from basic needs such as eating and sleeping up to the need for self-fulfilment (Maslow, 1943). Transport itself is a result of needs that can't be met in situ (Becker *et al.*, 1999; Gerike, 2007). To meet these needs, people or goods must cover distances, which involves interaction between individuals (attitudes, perceptions) and the accessibility of places (their distribution across space, as well as infrastructure). Hägerstrand (1970) was one of the first researchers to explain these interactions in a spatial-temporal context, writing as he was towards the start of the quantitative revolution in geography. To describe the duality of what he terms constraints and possibilities, he distinguishes between three types of constraints: a) individual temporal or *coupling constraints* from given appointments and resultant available times; b) spatial or *capability constraints* from accessibility of locations within the available time; and c) additional spatial or *authority constraints* from admission restrictions like private property or – together with temporal constraints – opening hours.

By adding individual socio-demographic characteristics, Geurs and van Wee (2004) classify these accessibility attributes into four different groups: a land-use component (locations and characteristics of opportunities and of demand), a transport component (location and characteristics of infrastructure for passenger and freight travel), a temporal component (opening hours of shops, available time for activities), and an individual component (income, gender, educational level, mobility resources such as vehicle ownership, etc.). Individual mobility decisions are made within this context, eventually leading to physical movement and therefore becoming measurable. Lanzendorf and Scheiner (2004) extend this model by including social and psychological aspects. They define a framework of structural (spatial and urban structure, transport system, temporal structures, economic factors, political and planning factors) and individual social and psychological attributes (demographic and socio-economic factors, social situation, lifestyles and mobility styles, attitudes and norms, availability and ownership of means of transport) as basic conditions for the genesis of travel.

Accordingly, travel behaviour is both complex and multi-faceted, but has largely been explored through reference to three key components, which are:

*   Travel purpose, often classified by specific activities such as work, shopping, taking young people to school, leisure trips;
*   Travel mode, which usually includes the dominant mode within a journey, but might also include secondary options;
*   Journey distance and/or time.

It is clear from exploring these attributes that measuring travel behaviour can be a complex process and evidently any notion of travel behaviour must be seen within the context of both individual and structural factors that can have a role in determining decisions to travel. Accordingly, a deeper understanding of psychological factors for travel decisions is a necessary basis for successful measures to influence behaviour. For many years, transportation research has relied on approaches using the model of a 'homo economicus' (Persky, 1995), implying that the main criterion for individual decisions is the maximisation of benefits. Such an economically-based rationale is grounded in the logical positivist tradition that we discussed earlier in this chapter and a direct application to travel behaviour is often made by using 'Rational Choice' theory (e.g. Simon, 1955). As an example, Gorr (1997) sees the attractiveness of a means of transport – defined by travel time, travel costs and quality – as the main factor for travel behaviour decisions (Heggie, 1978; Mahmassani and Jou, 2000; Sirakaya and Woodside, 2005). This economic rationalism is characteristic of approaches adopted in policy-making within nation states such as the UK, where the civil service is dominated by those with training in quantitative social sciences. This means that to make arguments other than those based on scientifically verifiable evidence, in the traditional sense, is very hard and therefore tends to drive down decision-making to a question of how it is possible to quantify 'factors' and 'barriers' that limit behaviour, and what the economic cost-benefit analysis is of adopting particular options.

One of the main criticisms of this dominant approach is founded on a critique of the rationality of human choices and argues that because of different limiting factors like incomplete information or influence from emotions (Frank, 1990) people are only able to make subjective decisions. Another point of critique deals with the mainly individualistic nature of Rational Choice Theory, hindering the identification of social norms and practices (Green and Shapiro, 1994; Sen, 1982). To take account of these critiques, travel behaviour researchers have attempted to adopt existing psychological constructs and theories within their research to permit a higher level of complexity. One such popular example, which we shall come across again later in this chapter and in Chapter 8 is the Theory of Planned Behaviour (TPB) (Ajzen, 1991). Like Rational Choice Theory, the TPB focuses on a maximisation of the satisfaction of needs as the basis for decisions. However, this concept is widened by including social norms and rules and accounting for subjective perception. Hence, the Theory of Planned Behaviour allows a more detailed view on travel behaviour and mobility decisions, and many researchers use this approach to get insights into the nature and structure of travel choices (e.g. Anable, 2005; Bamberg and Schmidt, 1998; Hunecke *et al.*, 2007).

Based on this theoretical approach, researchers are able to delve into greater detail concerning what has become a key focus for travel behaviour researchers in the last 20 years or so – the attention paid to the role of habits (e.g. Gärling and Axhausen, 2003; Kurz *et al.*, 2015). Møller (2002) gives a detailed overview on the role of habits for travel behaviour and mode choice. Verplanken and Aarts (1999, p.104) describe habits as 'learned sequences of acts that have become automatic responses to specific cues, and are functional in obtaining certain goals or end-states'. This automation is the main reason for the common resistance of individual travel behaviour to influencing measures, especially under stable contextual conditions (Ouellette and Wood, 1998; Verplanken *et al.*, 1997). It has been argued that a habitual behaviour is therefore very functional for the individual, because it reduces the necessary mental activity and eases complicated behavioural processes in stressful situations (Verplanken and Aarts, 1999) and allows individuals to handle multiple tasks (Ouellette and Wood, 1998). In this way, a habitual behaviour is established when a deliberate decision is made and frequently repeated in a stable, supporting context, and when this decision always leads to a satisfying result (Møller, 2002). In regard to travel behaviour research, the term 'state dependence' (e.g. Dargay and Hanley, 2007) has been applied to habitual behaviour because it describes the dependency of travel behaviour on learned actions, which reinforce particular practices and often means that the process of establishing new and changed habits is problematic. This reinforcement of habits is likely to intensify an individual's focus on one chosen means of transport and reduces the perception of travel alternatives (Fujii *et al.*, 2001; Verplanken *et al.*, 1997). Additionally, habits tend to reinforce what is regarded as a biased perception of other means of transport, resulting in an increase in perceived costs; for example, Kenyon and Lyons (2003) showed that car drivers normally overestimate costs and travel times for public transport use. Both of these issues – selective and biased perception of alternative means of transport – have to be considered in attempts to change individual travel behaviour in general, and particularly for measures to influence travel behaviour towards a more sustainable mobility.

Travel behaviour research: some examples

The *modus operandi* of travel behaviour research has, without doubt, been the use of travel surveys, either on paper, as diaries or now online (Barr and Prillwitz, 2014). Such surveys deploy quantitative methods to provide numerically based data for statistical analysis, which ultimately lead to the verification or otherwise of theoretical models. In the following examples, we examine the ways in which researchers have used travel surveys to explore key psychological attributes.

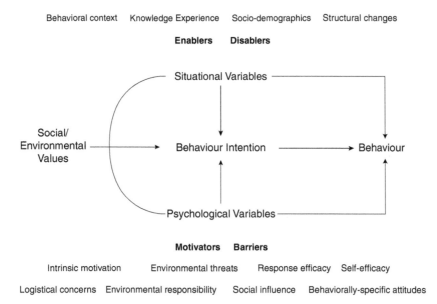

Behavioral context   Knowledge Experience   Socio-demographics   Structural changes

Enablers   Disablers

Motivators   Barriers

Intrinsic motivation   Environmental threats   Response efficacy   Self-efficacy

Logistical concerns   Environmental responsibility   Social influence   Behaviorally-specific attitudes

Figure 5.1. Barr and Gilg's (2006) conceptual framework of environmental behaviour.
Source: Barr and Gilg, 2006.

First, Prillwitz and Barr (2011) utilised quantitative travel-behaviour research to explore sustainable travel behaviour using a conceptual framework of pro-environmental behaviour, based on psychological research (Figure 5.1). In this research, the authors used Barr and Gilg's (2006) four key determinants of pro-environmental behaviour (values, situational factors, psychological factors and behavioural intentions) to characterise respondents into segments or clusters. The aim of this research was to explore the viability of segmenting populations so that particular interventions could be designed to suit the needs of particular clusters.

The travel-behaviour survey used in the research can be seen in Figure 5.2 and comprised a series of tick-box questions related to travel behaviour for different purposes, attitudinal attributes drawn from the literature concerning car use, walking/cycling and public transport use, alongside a series of questions regarding environmental values and socio-demographic items. These data were then analysed using cluster analysis, which is a multi-variate statistical technique for identifying groups within complex datasets, according to the similarity between individual respondents. The results of these analyses, in which four groups of daily travellers were defined, had the following varying attributes:

- Cluster 1: Addicted Car Users (n = 383). These individuals tended to use the car as their main mode of travel for most journeys and to live in low-density and commuter settlements. They tended not to hold pro-environmental attitudes and were politically conservative;

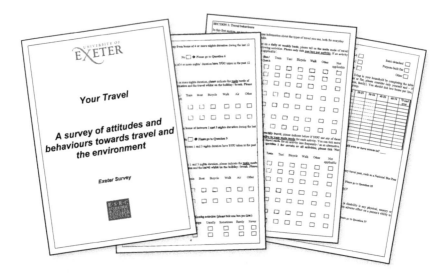

Figure 5.2. An example of a travel-behaviour survey.
Source: Authors.

- Cluster 2: Aspiring Green Travellers (n = 524). These were individuals who held relatively strong pro-environmental attitudes but who still largely relied on the car, although they would use alternative modes, especially walking, when possible. They tended to live in lower density environments and to come from a middle-to-high occupational background;
- Cluster 3: Reluctant Public Transport Users (n = 135). Although this group may appear 'green' from a behavioural perspective, they tended to hold fairly negative pro-environmental attitudes and in some instances were positive about car use. They tended to travel by public transport for several types of journey, although they may not have viewed this as an 'environmental' behaviour. They tended to be from older, retired groups who had less access to private motor transport;
- Cluster 4: Committed Green Travellers (n = 113). These individuals tended to be very pro-environmental in their attitudes and had relatively low levels of car use, relying also on walking and cycling (rather than public transport). They were the most politically liberal and 'green' and tended to come from largely professional and managerial occupations.

The findings from this research have subsequently been used to develop policy interventions to encourage individuals representative of the Aspiring Green Travellers segment to change their behaviour, on the basis that this group held pro-environmental attitudes but did not currently engage in a wide range of sustainable travel practices. Accordingly, what the segmentation analysis enabled was an exploration of key attributes that could be used to generalise about the broad trajectory of attitudes and behaviours of a group representative of a group in society.

A similar example can be seen from Anable's (2005) research into car and non-car users, which also utilised a segmentation approach. Anable's (2005) study adopted the Theory of Planned Behaviour (Ajzen, 1991) to explore the motivations of visitors to a National Trust site in the north of England, in which she measured psychological constructs relating to participant attitudes towards car-driving, subjective norms for car-driving, the perceived behavioural control felt by participants to change their behaviour, alongside behavioural intention and report behaviour. Her study generated six segments, four of which were car users and two non-car visitors. Once again, this research highlighted the ways in which representative samples from general populations can be utilised to produce generalisable trends for understanding the broad motivations and barriers for individuals to engage in particular types of travel behaviour.

Travel-behaviour research has therefore attempted to provide reliable, quantifiable and comparable data to understand travel choices, using a positivist and individualistic framework for exploring decision-making. Indeed, this kind of approach has been popular for many decades because it has been so closely related to the needs and demands of policy-makers, who wish to understand the quantifiable nature of travel behaviour and how behaviours, notably habits, can be broken and changed in response to anthropogenic climate change and the need to reduce reliance on car-based mobility. A useful illustration of the way in which segmentation has been utilised is given by the UK Department for Transport's (Department for Transport, 2011) segmentation model, designed to explore how personal travel choices could be amended through targeting particular interventions. Informed by academic research by scholars such as Anable (2005), Department for Transport's (2011) segmentation model identifies nine segments in the UK population, which are described in Figure 5.3. These segments clearly identify key groups who, according to the parameters of the model, have high travel needs and are likely to be having a disproportionate impact on carbon emissions. For example, both the car-owning 'Affluent Empty Nesters' and 'Educated Suburban Families' have high levels of travel need, but also hold positive attitudes towards the environment. From the perspective of targeting policy and allocating resources for promoting change, it is likely that these groups represent better opportunities for change than, for example, those in the 'Town and Heavy Rural Car Use', who are less predisposed to be positive about messages to reduce carbon emissions. Indeed, these findings are very much in line with Barr and Prillwitz (2011) in their research on daily and holiday travel behaviour, which are discussed in more detail in Chapter 8. Moreover, it is interesting to note that the three non-car owning segments largely have no access to private motor transport because of their age or residential circumstances, rather than a commitment to reducing environmental impacts.

As we will explore in Chapter 8, the use of quantitative segmentation studies, many of which are grounded in a social-psychological tradition, has had a profound impact on the engagement some academic researchers have had with

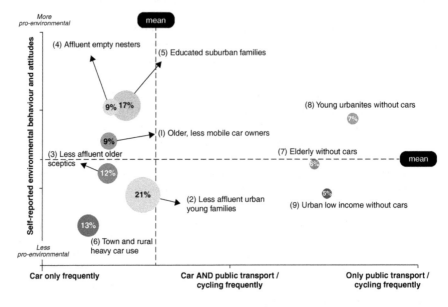

Figure 5.3. The UK Department for Transport's (2011) personal daily travel segmentation
           model.

Source: DfT, 2011.

central government in countries like the UK. The major shift towards focusing on behavioural change in policy (French *et al.*, 2009; Thaler and Sunstein,
2008) has meant that the work of quantitative social scientists, including travel
behaviour researchers, has become increasingly valuable to a political community seeking to change individual behaviours to achieve policy goals. However,
as we will note in Chapter 8, the complexity of behavioural change and questions underpinning the basic epistemology of behavioural research has led to
calls among policy makers for other social scientists, in particular those utilising
the new mobilities paradigm, to have a greater engagement with policy. It is to
the ways in which mobilities researchers have explored daily mobility that we
therefore now turn.

Mobility in daily life

Founded on substantially different epistemological assumptions, mobilities
researchers have sought to change the approach in transport geography away from
a focus on the individual and discrete (and quantitatively measurable) behaviours,
towards analyses of social practice and lived experience. At the most basic level,
the questions being asked by mobilities researchers about everyday travel are different; mobilities researchers are interested in how people experience mobility
and what this can tell us about social relations and the (historically and culturally

rooted) ways in which mobility practices develop. The approach therefore draws strongly on the agendas set out by Sheller and Urry (2006) and Urry (2007), and combines the contexts in which Cresswell and Merriman (2011) have argued we should be working alongside: practices, spaces and subjects. Analyses of daily mobility are therefore about adopting an interpretative epistemological framework that has a range of characteristics, which can be summarised using both mobilities literature (e.g. Fincham *et al.*, 2010; Spinney, 2009) and social science readers (e.g. Bryman, 2011).

First, studies of daily mobility have largely been drawn from a sociological tradition, where the focus is largely on the exploration of social, rather than individual, phenomena. As such, rather than deploying grand narratives or extant theories of individual behaviour, which can ultimately be 'tested', sociological approaches often seek to derive grounded theories of social practices and change. There are of course variants, but essentially sociologists and mobilities researchers try to work with data to interpret a theoretical narrative from the ground up, rather than formulating a generalisable theory from the top down.

Second, mobilities researchers, in adopting an intensive and interpretivist perspective on daily mobility are not concerned with generalisable data at the population scale and so are not concerned about probabilistic representations of the general population. Rather, what's of concern are the deep, meaningful and rich experiences of participants in research and what this can reveal about the contexts that surround them. As such, researchers of everyday mobility have tended to advocate qualitative, ethnographic and 'mobile' methods (e.g. Spinney, 2009) to capture mobility literally in motion.

Third, mobilities researchers recognise their own place in both the process of research and also in embodying mobility. Rather than trying to remain abstracted from the research process, researchers have assumed that mobilities, like all other social practices, are subjective in nature and that there is much to be gained from recognising the embodied experiences, emotions and sometimes pains of being mobile (e.g. Wylie, 2009).

Finally, to reinforce the point, mobilities research is about going beyond the individual. We noted in regard to travel behaviour research that individualistic concerns have dominated the social scientific of understanding daily travel for many decades, but mobilities researchers are far more concerned with how daily mobility for the individual is influenced by and changes the mobility of others and how such mobilities form social practices through interactions with spaces and contexts. Indeed, some within social science (e.g. Shove, 2003, 2010; Shove *et al.*, 2012) have argued that the key to unlocking our apparent dependence on carbon-intensive practices and seeming unwillingness to change our individual behaviours needs to be looked at through the lens of a socially focused analysis. Our practices are complex interactions involving individual cognition, material relations, social interactions, economic and political contexts and the different spaces these occur in. As such, to understand daily mobility is to appreciate complexity and context. Indeed, perhaps it is this realisation that makes the challenge of trying to change mobility practices, like all others, so much of a 'wicked' problem (Head, 2008).

Daily mobilities: mobile methods

Appreciating the complexities of daily mobility means that, unlike travel behaviour research, there are fewer commonalities between the approaches adopted to study everyday travel. For example, Jain (2009) utilises auto-ethnography to characterise the performance of bus travel in urban contexts through a series of 'acts'; Vannini *et al.* (2009) utilise ethnography to capture the experience of young people experiencing ferry travel; Laurier and Dant (2012) used video ethnography to collect over 100 hours of data on driving practices as a way of exploring the potential for driverless cars; and Edensor (2011) explored daily commuting through in-depth interviews. These and many other examples illustrate that daily mobility is situated in time and space and holds a great deal of meaning for both individuals and as a way of participating in society.

A useful example of how such research has had a major impact on geography in particular, and for which there is much potential in terms of policy relevance, is Justin Spinney's (2009, 2010, 2011) scholarship on cycling. This research is important because it takes an issue that has firmly been the preserve of travel behaviour research, in terms of a basic assumption that encouraging more cycling is one major way of reducing carbon emissions and promoting sustainability. Indeed, what Spinney has brought to the issue is a theoretical and methodological criticality that enables us to appreciate the socio-spatial complexity and conflicts that surround cycling and the ways in which we can research it in motion.

Spinney's (2009) call for research into cycling is based on a familiar but nonetheless important cornerstone of arguments put forward by mobilities researchers, namely that:

> Conceptualisations of movement and mobility within geography are increasingly complicating reductive and sedentrist understandings that have tended to theorise mobility either as meaningless, or as the practical outcome of 'rational' decision-makers … [There is a need for] research into cycling to explore the content of the line between A and B in order to highlight the often fleeting and ephemeral meanings that can contribute significantly to what movement means. An essential part of this project is for research to focus on the 'immaterial' embodied and sensory aspects of mobility that have previously been neglected or marginalised.
>
> (2009, p. 817)

In this way, Spinney challenges conventional approaches to studying cycling in geography that have tended to reduce cycling's value to merely a utilitarian means of reducing carbon emissions, alongside a tendency to treat cycling as a practice that is dis-connected from all other. In making this argument, Spinney urges geographers to consider the act of mobility, of being in motion, as something which has value and which cannot be explored through using conventional research methods such as questionnaires, static interviews or static photography 'All forms of mobility are profoundly embodied and consequently much of the experience of moving has remained stubbornly beyond the means of the visual and the verbal to decode' (Spinney, 2009, p. 818).

Using these methods, Spinney argues, will enable researchers to plot the complex relationships between the body, technology and space, thus demonstrating 'the limits to a research agenda that conceives of mobility as a rationalised and instrumental practice' (2009, p. 818).

In advocating a new research agenda, Spinney contends that new possibilities of understanding practices, spaces and subjects could explore the roles of technologies of cycling; bike activism; urban cycling; road racing; mediatisation of cycling; and cycling and identity. Such an approach has value because 'the bicycle [...] assumes a liminal status in the urban environment because it is more than simply transport; it is many things to many people' (Spinney, 2009, p. 826). In this way, the dual benefits of theoretical and methodological innovation can be achieved through adopting a radically different approach to cycling and, by extension, to other practices.

Accordingly, Spinney's exploration of cycling raises the scope for applying a range of mobile methods (Fincham *et al.*, 2010) for exploring daily mobilities. Indeed, what these and other interventions bring to the debate about daily travel is the way in which mobility is deeply related to other social practices and how the act of being mobile is profoundly experiential and can often be contested. For example, recent media stories about the dangers of cycling in London (e.g. BBC, 2015) have highlighted the tragic consequences of conflict between road users, which has led to cycling and the tensions with motor vehicle users becoming a politicised issue. Indeed, for geographers, these tensions and conflicts raise important questions about how spaces are configured and the ways in which 'shared space' for cyclists and pedestrians becomes contested.

Conclusion: travel behaviour, daily mobility and policy

The epistemological and methodological divide that we outlined in Chapter 4 between conventional transport geography and mobilities approaches is perhaps at its starkest when we examine the issue of daily travel, because it is where a long heritage of travel behaviour research has been challenged by alternative readings of the everyday. As we noted in Chapter 4, the opportunities for linking these two very different approaches are limited because of the epistemologies involved. Indeed, to some extent, the two approaches seek to address different questions: travel behaviour research is interested in understanding how individuals make travel decisions and what consequences this has on networks and systems; mobilities researchers are interested in appreciating the experience and value of mobility, and how this relates to other social practices. As a consequence, travel behaviour research has often been regarded as more familiar territory to those within government because of the shared assumptions of behavioural economics and the adoption of the scientific methods to understand society. Yet it is also the case that decades of behavioural research from one perspective, be it on travel, recycling or energy use, has only partially uncovered the complexity of human behaviours and has come only a little way towards understanding how behaviours can be changed.

This is perhaps where a mobilities approach can have an important impact outside of the academic community (and therefore recognising the importance of not simply viewing research in an instrument of policy). At its heart, the new mobilities paradigm argues that travel is a social process and one that reflects the wider socio-economic circumstances of the spaces in which it takes place. As research by Barr and Prillwitz (2014) highlights, these contexts can be ones that have developed over generations, resulting in contemporary social practices. They use the example from one of their research participants, who highlighted the change in working cultures in the UK from the 1950s to early 2000s. In the immediate post-war era the UK was still largely a manufacturing-based economy, with large numbers in the workforce employed in factories, with company-built properties nearby and highly regularised and routinised working days. Mobility was therefore highly regulated and determined by relatively few employers, who often provided the infrastructure to support mass travel to work. In contrast, the move to a service economy, with growth in small businesses and a resultant spacing-out of workplaces has promoted greater need for private mobility based on motor transport. Indeed, these economic changes are clearly reflected in the broader political shifts in countries like the UK, where a focus on neo-liberalisation of the economy has promoted individual choice as a cornerstone of modern democratic society (Clarke et al., 2007; Giddens, 1991). As a result, the practices we see today as 'simply' individual travel-mode choices are, in part, a reflection of the structure of society and economic activity. From the perspective of a changing climate and the need to reduce carbon emissions and promote local residence, this core message of the new mobilities paradigm is one that should not be ignored.

There are major challenges for viewing daily travel from this perspective, as well as one in part influenced by cognition and individual decision-making. This is because we need to look both at and beyond the individual and do so with an open methodological perspective. Intellectually, this partly becomes an issue related to language; 'behaviour' has been pitted against 'practice', yet we should be able to work with both of these terms in the common pursuit of change (Wilson and Chatterton, 2011). Moreover, finding common ground will inevitably mean some movement on either side of the debate. Politically, there is also the question of how academic researchers can engage policy-makers and elected politicians in questions that are long-term and challenge existing economic orthodoxies. Behavioural change, as shall see in Chapter 8, has provided a clear pathway for advocating a neo-liberal approach towards governing the environment (Jones *et al.*, 2011a, 2011b), but it is likely that more fundamental shifts will be required to see major shifts in mobility practices that could viably reduce carbon emissions. If we only see the issue as being about the individual, we are unlikely to make more than token progress.

6 Consuming places
Leisure travel and the 'end of tourism'

Introduction

Travel for leisure and tourism is a global phenomenon that continues to increase in economic and social terms. International tourist arrivals grew globally by 5 per cent in 2013 to 1,087m from 1,035m in 2012 (United Nations World Tourism Organization (UNWTO), 2015). Moreover, in terms of leisure-travel spend, the World Travel and Tourism Council (2016) estimates that travel and tourism grew by 3.1 per cent in 2015. This growth supports the claims that tourism and travel is one of the world's largest industries worth almost $7bn. In other ways such figures represent a complex set of flows or as Urry (2000) stated tourists 'dwell in various mobilities' (p.157). Such mobilities are surrounded by and part of global production systems. Tourism and the related ideas of mobilities are part 'of the same set of complex and inter-connected systems, each reproducing each other' (Sheller and Urry, 2004, p.5). The aim of this chapter is to explore these systems in the context of both tourist behaviour towards travel and the systems of tourism mobilities. In developing these ideas, we call on a number of perspectives including studies associated with the behavioural aspects of leisure, travel and tourism, through the lens of mobilities. In the latter context, as Sheller and Urry (2004) explained, tourism itself is not just an important form of mobility but it is both informed by and informs mobilities. Hannam *et al.* (2014) go further arguing that: 'travel itself not only provides the means of tourism activities but may also be a key experiential feature for many tourists'. Moreover, the concept of auto-mobilities helps provide a much broader set of ideas which we will explore later in the chapter. Related to these ideas are the notions of experiences created in tourism mobilities through the use of mobile technologies (Hannam *et al.*, 2014).

We start with a short discussion of the evolution of tourism travel through the lens of the growth of popular travel and the innovation of the package holiday, along with an examination of the notion of auto-mobility in a leisure-travel context. This is followed by a much more detailed discussion of tourism and mobilities through the lens of tourist behaviour. Finally, attention will be given to the creation of what Edensor terms (2007) 'touristscapes', which are reproduced via the 'growth of tourism reflexivity' (Sheller and Urry, 2004, p. 3). This in turn is conditioned by many forces including public institutions, global capital and international organisations. Such institutionalisation is framed within a network of media, selling travel and tourism, and increasingly the co-creation of travel experiences. Finally, we conclude by examining the ideas and growth of slow tourism in terms of its impact on travel.

The rise of the tourist class and the growth of mobility

During the aftermath of the Second World War (post 1945) there arose increased opportunities across Europe and North America for leisure travel, giving rise to a new tourist class. These people were not the wealthy of the nineteenth-century travellers (Towner, 1996) or indeed the day visitors to seaside resorts of the early twentieth century (Walton, 1983). This new tourism class was being offered exciting opportunities for international travel facilitated in part by increased paid holidays (in the UK this was the Holidays with Pay Act of 1938 which was only effective after the end of the Second World War (Williams, 1998)), and a growing level of car ownership (Berghoff *et al.*, 2002; O'Connell, 1998). Existing discussions of this rise in leisure mobilities have focused on the development of the package holiday with what Walton (2011) has criticised as an over-emphasis on the air-inclusive package. For example, Wright (2002) in her discussion of the rise of package holidays among UK consumers mentions only those relating to air travel, ignoring the importance of trips by coach, and, of equal importance, the impact on tourist travel.

All of these earlier developments were associated with a series of related innovations providing new ways of travel and enabling the British holidaymaker to cope with foreign travel. These included the development of easy credit facilities in the late 1960s and early 1970s before the availability of credit cards. For example, the British-based holiday company Wallace Arnold, developed a holiday budget plan, in their words: 'to make it easy for you to afford the holiday you deserve' (Wallace Arnold, 1970, p. 1). In much of their information the complexities of international travel were hinted at in the brochures but most importantly Wallace Arnold offered solutions and an advice service. By the early 1970s the same company was highlighting their experience in holiday planning. Their brochures proudly claimed: 'From the moment you book to the moment you return, our whole organization is geared to ensure maximum pleasure for you' (Wallace Arnold and Wallace Arnold Air, 1972, p. 3). The emphasis on modernity was reinforced by their claim to be one of the first travel companies to computerise their reservation and administration system (p. 2). Such reassurances were seen as necessary given the views of would-be travellers in a survey undertaken in London during 1961. This found that only 11 per cent of the 1,000 people interviewed took a holiday abroad, and the main reasons for not doing so related to cost and psychological factors, i.e. lack of confidence (Associated Rediffusion, 1962). This level of uncertainty with foreign travel is identified by one of the holidaymakers who had travelled with a package holiday company, Gaytours based in Manchester. The respondent is talking about a holiday taken in 1963:

> This being our first time abroad we were a little apprehensive but I now say in all honesty that this was needless. The hotel accommodation and food were first class and the services of your representatives left nothing to be desired.
>
> (Gaytours Brochure, 1964, p. 53)

The reaction of consumers to these developments was tangible as shown in the rapid increase in those widening their travel horizons and in the UK the number of holidaymakers travelling abroad increased from 1.5m in 1951 to 8m in 1975 and by 1987 had risen to 20m of which over 50 per cent were holidaying using inclusive tours (British National Travel Survey, 1976; Horner, 1991). The growth of foreign holidays contributed to new patterns of mobility and the development of Fordist patterns of consumption and production based around the package holiday (Shaw and Williams, 2004). For example, the number of package holidays by air sold in the UK increased from one to two million between 1965 and 1970 (Burkart and Medlik, 1981). By the late 1970s, the six major tour operators in the UK accounted for over 75% of inclusive packages, with Thomson's alone handling one million passengers per annum (Pearce, 1987). The essence of such consumption patterns is that of mass consumption related to large numbers of tourists associated with a circuit of mass production. This production in turn is highly dependent on scale economies (Ioannides and Debbage, 1998), although products tend to be differentiated along cost and on the basis of stage in family life-cycle segments.

Changing patterns of tourism consumption

In what we will come to examine as a generalised shift in tourist practices, patterns of consumption in tourism have gradually shifted since the 1970s from forms of mass tourism to those characterised by flexibility, including the creation of more specialised, individual niche markets. As we shall see later in this section, such shifts represent a move from Fordist modes for the production and consumption of tourist products and services (based on low levels of choice and variety) to the post-Fordist formation of niche markets (for particular segments) and finally the neo-Fordist trends for mass customisation in the twenty-first century. The tourists identified with such forms of consumption have been identified in a variety of ways (Mowforth and Munt, 2015) and the crux of these ideas concerns the timing of shifts in tourism consumption and their impact on travel behaviour. Such debates in tourism research relate to broader trends in consumption and consumerism studies (Miles, 1998; Slater, 1997) and in terms of the UK, Lee argues that a transformation took place in the 1980s, when the aesthetics of style and consumption became more diverse in response to an increasingly sophisticated consumer market. Of key importance was the growth of the so-called 'new' middle classes (Urry, 1990; Mowforth and Munt, 2003). These authors draw from Bourdieu (1984), who argued that different social classes are engaged in a struggle to distinguish themselves from one another, in part through consumption. As Munt (1994) points out, such consumption includes: 'objects and experiences such as holidays' (1994, p. 105). However, as Mowforth and Munt (1998) argue: 'data on the increased importance of new forms of tourism are difficult to come by' (1998, p. 98). Shaw (2005) has attempted to reconstruct the changes in the UK highlighting the move from Fordism through to post-Fordism and beyond (Table 6.1). As can be seen, only general trends are given but even so these are suggestive of important changes in both tourism and travel patterns. Before considering these it is worthwhile identifying some broad consumer trends that have impacts on tourism and trends that have impacts on tourism and travel behaviour. In part such changes relate to general discussions surrounding post-Fordist and neo-Fordist forms of consumption. To understand how such changes in consumption have impacted on tourism and travel it is in part necessary to examine the underlying processes in more detail. We do this by using examples from the UK in attempt to flesh out Table 6.1. Some of these underlying factors relate to broader shifts in consumer lifestyles that provide a link to the consumption of holidays and mobilities. There are two dimensions to this, the first relates to consumer attitudes while the second is associated with more socio-demographic factors.

Changes in consumer attitudes involve an increase in the pursuit of individuality, more emphasis on informality and spontaneity, an increased use of all senses for personal well-being, and a greater willingness to integrate IT and the web into the overall consumption process (Kamarulzaman, 2007). Since the 1990s there has also been an increase in sustainable consumption patterns. In this context Soron (2010) sees sustainable consumption as an individual self-identity construct.

Table 6.1. General shifts in holiday consumption patterns in the UK and related travel

Phase	Characteristics	Travel
Fordist 1950s–1970s	Development of inclusive packages. Little variety, price driven, increasing interest in market segments based around family life cycle	1950s coach travel 1960s inclusive air packages, more long-haul flights
Post-Fordist Mid-1980 1990s	Increasing interest in heritage tourism from 'new middle classes' Greater demand for ecotourism holidays	Increase in long-haul flights Dominance of air travel
Late 1990s	Interest in more independent travel	Impact of low-cost airlines
Neo-Fordist 2000 to present	• Twenty-first-century impact of mass customisation • Consumers have more control • Growing interest in slow travel	Specialist walking, cycling holidays

Source: Modified from Shaw (2005).

This fits with the ideas of the development of a form of cultural capital by many consumers, who attempt to adopt sustainable lifestyles (Seyfang, 2005).

A second group of factors impacting on consumption patterns relating to tourism and leisure patterns are associated with demographics. There are two significant trends in many developed economies such as the UK, namely: an aging population along with the creation of more single-person households. The former is responsible for an increase in retired households with a more active generation of over sixties. The lengthening of life expectancy also means consumers have longer to develop as tourists and generate more leisure trips. At the other end of the age spectrum younger people are establishing single-person households and marrying or forming partnerships later in life. Single-person households increased from 18 per cent in 1971 to 29 per cent in 2003 but the figure fell to 19 per cent in 2013 due to the recession and affordability. These people form a significant segment in the increasingly important youth market. Many of these may also be younger professionals with complex lives linked to a strong sense of independence and control over their leisure and holiday experiences.

In contrast, work practices within restructured economies like the UK have impacted on consumer behaviour towards leisure. Those in employment have seen leisure time decline by 100 hours per year since the late 1990s. Not only are there longer working hours but there is a tendency towards less job security (Haworth and Lewis, 2005; Lyons *et al.*, 2007). This for some people has helped define an emphasis on a better quality of life, with a search for an easier more natural form of living; expressed in the types of holidays being taken. Evidence also points to work pressures impacting on holiday taking with some one in three managers in the UK not taking their full holiday entitlements. Similarly, in the USA surveys have revealed work constraints and financial concerns as main impacts on holiday travel (Lyons *et al.*, 2007; Petrick and Durko, 2013).

The other area of impact on holiday-travel behaviour relates to the changing cost of travel and the impact of online travel agents (Graham and Vowles, 2006). The increased growth of low-cost airlines has significantly reshaped the geography of leisure travel, especially within Europe (Graham and Vowles, 2006). The innovation of low-cost airlines is based on deregulation of airlines, the use of credit card bookings on the Internet and the lowering of operating costs where possible (Graf, 2005). The growth of operators such as Ryanair and EasyJet in the UK have provided fresh opportunities for cheaper leisure travel and opened up new destinations across Europe (Graham and Shaw, 2008). For example, passenger numbers from UK airports grew by an average of 6 per cent per year from the mid 1990s, while average air fares across Europe fell by 26 per cent between 1998 and 2003 (Njegovan, 2006). The reduction in air fares has altered the family budgets of leisure travel by air on short-haul routes. Survey data from a sample of 550 outbound passengers from the UK airport of Stansted showed that the share of air fares in total expenditure on foreign holidays is between 25 and 35 per cent of total expenditure (Njegovan, 2006), representing considerable savings compared with previous periods. Within Europe low-cost airlines have given savings of up to 50 per cent (Francis *et al.*, 2007). Work by Martinez-Garcia and Raya

(2008) in Spain has shown visiting tourists using low-cost airlines spend almost 33.8 per cent of their budgets on travel compared with 37.5 per cent for those using conventional airlines.

Technology and more especially the use of social media and the Internet have changed access to information and the holiday-planning process (Fotis *et al.*, 2012). In 2001 around 39 per cent of UK households were online and flights and travel information represented 31 per cent of usage accounting for 14 per cent of all UK travel and leisure sales (English Tourism Council, 2002). In 2013 69 per cent of UK households planning to holiday abroad planned to book this holiday online (TNS, 2013). As we shall see later social media is an increasingly significant factor in leisure travel planning (Yoo and Gretzel, 2011). Thus in a recent survey in the UK, 42 per cent of respondents claimed that sites such as Trip Advisor or Travel Supermarket provide them with major sources of information (Travel Supermarket, 2013). In terms of holiday bookings by UK consumers for holidays abroad 46 per cent still bought traditional package holidays compared with 35 per cent who booked travel and accommodation themselves with different companies, along with 20 per cent who booked separate travel and accommodation via the same company, while the remaining 23 per cent booked either travel only or accommodation only (ABTA Consumer Survey, 2014).

Leisure mobilities – flying around

The factors underlying leisure consumption outlined in the previous section have led to marked shifts in travel behaviour. In this part of the chapter we turn our attention to changes in leisure travel. The travel horizons of consumers in many mature markets such as the UK have increased from the 1970s onwards as more exotic long-haul destinations were on offer. Foreign holidays of UK residents have increased from 17.5m in 1980 to 58.5m in 2013 (House of Commons Papers, 2016). Such numbers reflect leisure holidays and include visiting friends and relatives. These increases have been driven largely by lower air fares along with more route connections. The influence of low-cost carriers has however spilled over into embedding holiday practices as an idealised type of travel freedom for many consumers (Becken, 2007; Shaw and Thomas, 2006). Burns and Bibbings (2009) have also highlighted how some cash-rich and time-poor consumers have been induced by the growth of low-cost carriers offering long-haul flights, with shopping trips to New York and city breaks to China. Most of such developments have grown in the twenty-first century, led by a variety of low-cost airlines (Wensveen and Leick, 2009).

The link between increased air travel and leisure consumption has been viewed as part of what Urry (2010) has termed 'high mobility systems' (2010, p. 88) or 'mobility complex' (2010, p. 90). This includes a number of components that in their totality remake consumption, pleasure, work, friendship and family life (Urry, 2010). Those that particularly impact on leisure mobilities include: the current state of movement around the world; the range of mobility systems in use; the growth of places that have to or may be travelled to; the increased capacity to compare and contrast places (via the Internet); and the increased significance of multiple mobilities for people's social and emotional lives. In terms of the last factor: 'touring the world is increasingly performed' (Urry, 2010, p. 91), which means the emotions of pleasure seeking and of novelty of experiences become of ever greater significance. These ideas link with the growth of what Pine and Gilmore (1999) have termed the experience economy, which we consider later in the chapter. As a consequence, air travel has switched from being an extraordinary event as it was for the mass tourists described earlier in the chapter, to what Randles and Mander (2009) see as the domain of the everyday. This has led authors such as Cohen *et al.* (2011) to identify a behavioural addiction to flying or 'binge flying'. This notion takes the ideas of addictive consumption into the realms of leisure travel. Cohen *et al.* (2011) identify such addictions, arguing that in the context of air travel 'long-term outlooks are sacrificed for immediate gratification' (2011, p. 1076). Table 6.2 shows some of the key characteristics of these so-called addictive behaviours from a range of studies relating to travel. The main thrust of such results concern the difficulties of changing behaviour and the different ways that frequent flyers tend to justify their behaviour. In this context, Shaw and Thomas (2006) have described air travel behaviour as a form of 'hyper-mobility', which they see in part as being socially aided by the technologies of social media (Fotis *et al.*, 2012). Others go further arguing that contemporary consumers are 'stratified by their degree of mobility – their freedom to choose where to be' (Bauman, 1998, p. 86, quoted in Shaw and Thomas, 2006, p. 211). In this context the views of consumers on air travel, regardless of the potential consequences on the environment, appear to be somewhat ignored. However, evidence suggests that many air travellers tend to justify their continual behaviour on the grounds that changing their travel patterns would have little effect on the overall impact on climate change (Barr *et al.*, 2010; Shaw and Thomas, 2006). Such views, however, are complex in that other wider survey data by the UK's Department of Transport (2008) have shown that 66 per cent of respondents believe air travel is harmful to the environment, with 44 per cent of this group mentioning climate change as an area of concern. This highlights not just an intention–behaviour gap but also the difficulties of consumers seeing a way to even reconcile their actual behaviour.

Table 6.2. Characteristics of addictive travel behaviour

Characteristics	Study (Author)
• Feelings of escape, experiences of pleasure and excitement, relaxation, disinhibition of behaviour	Rojek (1993); Ryan (2010)
• Unwillingness to reduce leisure holiday air travel	Barr *et al.* (2010)
• View holiday planning process as a heightened pleasure activity – a form of cognitive narrowing with attention just focused on immediate pleasure	Elliot (1994)
• Return from trip functions as an habitual trigger towards planning future travel	Randles and Mander (2009)
• Difficulties of changing flying behaviour justifies frequent holiday air travel	Scott *et al.* (2010)

Source: Developed from Scott *et al.* (2010).

Table 6.3. Major contributions to climate change and related mitigation measures

Activities	Mitigation Measures [1]
Home	
Heating, lighting, disposal of waste	Low energy technology, recycling
Travel	
Car travel	Switch to cycling and walking if possible
Flying	No measures suggested; inverted behaviour: 'my time'

Source: Modified from Hares *et al.* (2010)[1] and Barr *et al.* (2010).

Research based on more qualitative data from focus groups reveals a number of behavioural responses regarding holiday travel by air. First, when asked about measures to mitigate their travel against climate change, flying did not have any responses (Table 6.3). In contrast all the other major contributions to climate change mentioned in the focus groups did (Hares *et al.*, 2010). Results from another study, again based on focus groups, highlights that many travellers changed their behaviour while going on holiday from one of sustainable behaviour at home to one more centred around selfish values (Barr *et al.*, 2010). This inverted or flip-over behaviour is typical of many tourists both in terms of their travel patterns and at their holiday destination (Shaw and Williams, 2004).

Studies have also examined the so-called intention–behaviour gap in the context of consumers intending to be more sustainable but then finding reasons not to make such changes (Barr, 2004). The attitudes towards flying and pro-environmental behaviour suggest according to Adey *et al.* (2007) 'aero mobility is now embedded in the global fabric' (2007, p. 785) contributing to the intention–behaviour gap. Hares *et al.* (2010) found in their survey that flying was the third most recognised impact on climate change by respondents but in spite of this no one mentioned changing behaviour to address this impact.

Driving ambitions: auto-mobilities and leisure and tourism

If flying has become something of an addiction for some consumers in recent years, the car has been an ever-growing influence on leisure trips and holidays. The importance of the car goes beyond the freedom it gives individuals as it is also an iconic product with 'which people gain considerable status from its sign-values' (Lucas and Jones, 2009, p. 10). Featherstone (2004) and Urry (2007) have encapsulated such values of the car under the term 'auto mobility'. As Urry (2007) explained: 'the seamlessness of the car journey makes other modes of travel inflexible and fragmented' (2007, p. 119).

Featherstone (2004), along with Lucas and Jones (2009), have recognised three main phases relating to the car as a consumer object. We can use these also to categorise the car's relationships with leisure trips. The first phase (c.1900 to c.1925) is the era of large luxury cars for upper-class consumers and as such had little impact on leisure travel in a wider social context. Phase two (between c.1926 and c.1960) was the period of increasing mass production of cars along with changes in car ownership. In terms of leisure travel, it marked an increasing recognition of the freedom the car gave. New magazines such as *Autocar* (1929) proclaimed: 'public transport, no matter how fast and comfortable, inflicts a sensation of serfdom which is intolerable to a Free Briton' (quoted in O'Connell, 1998, p. 79). This freedom was also soon to be harvested into what contemporary writers such as Morton (1935) in his influential book, *In Search of England*, gave a picture to the would-be motorist of romantic adventures along the roads of England. Such travel guides took on greater significance for motorists with the publication of *The Shell Guides to the Counties of England* from 1934 onwards (Knights, 2006). The latter parts of this period also corresponded to the growth of package holidays and mass holidaymaking.

The third phase, c.1960 to the present, is seen by Featherstone (2004) as one where the car is 'part of a fragmented series of subcultures in which a whole range of new types of vehicles emerge targeted towards small niche markets' (Lucas and Jones, 2009, p. 11), which focus on the identities and consumer cultures. In terms of leisure trips, the popularity in the early parts of this third phase became an important trend as car ownership increased in the UK from the 1960s. The number of British households with a car by the mid 1970s rose to 59 per cent increasing further to 77 per cent by the mid 1990s and 80 per cent in 2013 (Department for Transport, 2014). Of equal significance has been the number of females who hold a full driving licence, rising from just 29 per cent in the mid 1970s compared with 69 per cent of males. By 2013 the comparative figures had changed to 66 per cent of females and 80 per cent of males (Department for Transport, 2014). This democratisation across the gender divide has been another factor affecting both the number and types of leisure trips taken by car.

It is within this current third phase that the ideas of auto-mobility are more widely felt. In terms of leisure behaviour Urry (2007) claims that this phase has seen the rise of 'free and independent travellers' (2007, p. 121), leading to a type of post-Fordist consumption. Hannam *et al.* (2014) has drawn together a number of strands that underlie the importance of the car to leisure and holiday travel. These fall into three main headings, namely: freedom and adventure; co-existence in private, controlled spaces; and inhibiting interactions.

In terms of the car inducing positive feelings of control and freedom, a number of authors have stressed the notion of touristic adventure. Edensor (2007) sets the importance of the car in terms of 'linkages between auto mobility and national identity' (2004, p. 103). Such linkages include broad-based but rather limited evidence about the nature of identity and leisure trips in this context. Indeed, many of the studies relating to aspects of auto-mobility in a leisure setting are rather limited to narrative accounts from ethnographic studies. Huijbens and Benediktsson (2007) provide an ethnographic study of auto-mobility within the interior of Iceland, showing how the arrival of the jeep opened up travels to the more rugged highlands. The narratives within their study highlight the complexity of travel through such landscapes in terms of the traveller's emotions.

The car, therefore, gives a notion of tourist adventure allowing the occupants to travel through a variety of landscapes and in doing so provides the users with the 'cultural dreams of adventure' (Hannam *et al.*, 2014, p. 176). Such adventures are made easier through sat-nav systems that guide the driver through unfamiliar places. Mobile technology adds to these experiences by allowing the tourist to communicate with others who are not on the journey, both in words and pictures via smart phones. Bull (2004) goes further arguing that the soundscapes in terms of in-car entertainment and communication systems available within contemporary vehicles help enhance the feelings of adventure, transforming the spaces they pass through.

Hannam *et al.* (2014) raise however a note of caution, acknowledging 'the feelings of autonomy may also be hindered rather than advanced by the car' (2014, p. 176). Rather than dreams of adventure and release, cars have produced negative emotions, anger, envy and road rage, hindering notions of escapism and freedom (Beckmann, 2001). A further and very significant consequence of our love affair with the car is that it distorts aspects of rational behaviour. As Sheller (2004) points out 'car consumption is never simply about rational economic choices but is as much about aesthetic, emotional and sensory responses to driving' (2003, p. 2). This view also offers an alternative pathway that sees the distortion of rational-choice models which are used in many of the debates on transport planning and policy.

The car and leisure trip behaviours

The mobilities literature has brought a fresh perspective to our understanding of behaviour towards the use of the car. The majority of these studies have been of an ethnographic nature with Sheller (2004) calling for more qualitative work that 'will take into account how these apparently internal psychological dispositions and preferences are generated by collective cultural patterns' (2004, p. 223). These ideas of examining such behavioural drivers are important and in this part of our discussion we explore other behavioural perspectives on leisure travel. We start by briefly exploring some basic empirical information on cars and leisure travel using examples from the UK, before considering behavioural studies.

Within the UK 61 per cent of all leisure trips are undertaken by car and between 1995–1997 and 2013 the average trip length increased from 8.4 miles to 9.5 miles. Over the same time period the average number of trips fell by 15 per cent mainly due to a decrease in visiting friends and relatives. This is partly due to the frequent use of the Internet and social media especially Facebook. In contrast, day visits for leisure purposes have increased (Department for Transport, 2016b).

What these statistics fail to highlight is the growing complexity of many leisure trips, which increasingly have become more multi-purpose due to the freedom provided by the car. Other survey data on leisure behaviour shows the car accounts for 65 per cent of all day visits amounting to around 1.58m trips in 2014 (Visit Britain, 2016). These data highlight something of the complexity of activities behind such day trips (Table 6.4). As can be seen many activities are undertaken as part of what people consider to be a main recognisable activity, although in this survey almost 6 per cent of visits had no single main activity. Of course these data do not indicate those trips taking place by car. However, given the large proportion of leisure trips by car we can safely assume that the majority did take place by car. Therefore, to a great degree Table 6.4 reflects the flexibility in leisure behaviour facilitated in large part by the car.

The same degree of flexibility in terms of tourist travel behaviour has been noted by Tideswell and Faulkner (2002). While empirical evidence is somewhat scarce authors have argued that certain rhythms may define individual travel patterns (Dickinson *et al.*, 2011). More significantly, Zillinger's (2007) study of German car tourists travelling in Sweden gives an insight into the availability of time and to a degree spatial learning influences travel routes. In this study mobility was not significantly influenced by length of stay, but when tourist travelled to another place by car they usually did so 'to enjoy nature or to visit attractions or towns' (Zillinger, 2007, p. 78).

Other detailed empirical studies have highlighted the role of the car in terms of visits to heritage sites (Dickinson *et al.*, 2004) along with evaluations of trips to National Parks (Connell and Page, 2008). In terms of Dickinson's *et al.* (2004) study of historic properties and gardens owned by the UK charity, The National Trust, the main form of transport was by car, and the results of this research 'found that people travelled further on day trips compared with those on holiday' (2004, p. 108). They argued that having travelled to a destination on holiday such visitors tended to experience a form of travel inertia and did not want to make further long trips to attractions. Visitor behaviour in relation to National Parks has had more attention, although as Connell and Page (2008) explain the complexities of the car 'in shaping tourist patterns and destinations has not really been explored in any detail' (2008, p. 561). Their work in Scotland focused on the itineraries of different visitors along with an analysis of stopping behaviour. In this context they found that a large proportion of visitors were so-called 'one stoppers' (2008, p. 575) at key locations within the National Park.

The symbolic motives of car use in terms of leisure and holiday have been identified in a large-scale empirical study by Böhler *et al.* (2006) on case studies of three German cities; Augsburg, Bielefeld and Maddeburg. This survey of 1,919 people presents a large scale, detailed database of travel behaviour including key

Table 6.4. Volume of tourism day visits by type of activity in the UK, 2014 (volume in millions)

Activity	Main	Also Undertaken
Visiting friends and relatives	363	538
Going for a meal out	152	393
Undertaking outdoor leisure	122	255
Going on a night out	131	241
General day out	130	231
Special shopping	110	196
Going out for entertainment	96	142
Going to visitor attractions	77	125

Source: Visit Britain (2016). Note: in this survey leisure day visits are those lasting 3 hours or more.

Table 6.5. Modal split of holiday and short-stay trips in Germany

Type of Travel	Holidays/Trips (%)	Short-stay Trips (%)
Non-motorised travel	0.3	0.4
Trips by car	60.7	77.1
Public transport (localised) bus, tram, local trains	9.1	9.2
Public transport (long distance) trains	7.4	11.6
Air travel	22.5	1.7

Source: Böhler *et al.* (2006).

attitudes expressed towards the car concern, autonomy, privacy and experience. The mobility patterns of the sample showed that just over 387 trips/year were made for leisure and holiday purposes. In terms of the modal split between holidays and short-stay trips, Table 6.5 shows the dominance of the car, particularly as expected for short-stay breaks.

Factors affecting leisure and holiday travel trends

The study by Böhler *et al.* (2006) identified four holiday travel groups: group one, the 'non-traveller', mainly comprising older people and the unemployed. In other words, many were socially disadvantaged. They also have a low level of openness to change linked to both their low incomes as well as low interests in variety and new experiences. Group two, termed 'local travellers', had similar characteristics to non-travellers, but also comprised a number of married couples with young children. 'Mid-distance' travellers, group three, comprised a high percentage of two-income households, mainly middle-aged people with high incomes. Finally, 'long-haul travellers', group four, were characterised by highly educated young adults with no children, along with retirees, all of whom were high-income households. More importantly this group prefers more individualistic ways of living and travelling; indeed, they tend to assess the symbolic dimension of public transport as comparably low in importance: 'from Group 1 to Group 4 we find an increasing openness to change and self enhancement' (Böhler *et al.*, 2006, p. 660)

These findings in part fit with earlier, seminal work by Iso-Ahola (1983) showing that leisure travel provides various opportunities for consumers to engage in variety-seeking behaviours. Iso-Ahola (1983) saw this in terms of travellers attempting to satisfy underlying psychological needs, 'escaping' normal situations and 'seeking' more rewarding experiences. In the context of 'models' of leisure travel there is a need to bring together two areas of literature, namely that drawn from leisure studies of travel behaviour along with those from tourist travel. Smith *et al.* (2012) point out that these have 'been studied in their respective silos' (2012, p. 2207). Although a few authors have attempted to consider exploring what Carr (2002) terms the 'tourism-leisure behaviour continuum', they often have done so in a rather broad fashion usually ignoring patterns of travel behaviour. What does connect these two areas together in terms of leisure travel is the degree that travellers want to have 'novel' experiences or would rather have a high degree of the 'familiar'. Böhler *et al.* (2006) stressed this with regard to the four holiday travel groups. In more general terms of tourist behaviour early work by researchers such as Plog (1987) and Cohen (1972) identified the importance of 'novel' or 'familiar' experiences.

Attempts to understand or even model leisure travel including that by tourists has proved at best complex and somewhat limited. As Lew and McKercher (2008) point out even at the intra-destination level there are a wide range of variables that need to be considered, including destination characteristics and the socio-economic characteristics of tourists along with their motivations, interests and personalities. More recently the problems, though not resolved, have been addressed using different methods, which have included GPS tracking systems and the experimentation of applying a sequence alignment method to tourist travel behaviour (Shoval *et al.*, 2015). As the authors explain in the study of visitor movement around Hong Kong, sequence alignment methods are modified from biochemistry and in this instance used to construct sequences of the movements of tourists over time and space. The approach is still somewhat experimental but it does demonstrate the complexity of tourist leisure travel patterns at the destination level.

Many of the studies examining tourism and leisure travel have attempted to categorise or group people in terms of their characteristics and links with patterns of travel (Schlich *et al.*, 2007). The complexities of leisure travel, including tourist trips are increasing as demonstrated by Ettema and Schwaren (2012) who highlighted the trend 'towards more intensified and diversified forms of leisure undertaken more often outside the home and at a greater spatial scale' (2012, p. 175). The trends were based on lifestyle choices and the search for new experiences as identified in the ideas of Pine and Gilmore (1999) so-called 'experience economy'. For example, Schlick *et al.* (2007) have recorded the trend for people to continue to visit new locations for leisure activities in their study of travel patterns of individuals in Germany even over a six-week period. From this study and others (see Ettema and Schwaren, 2012, for a summary) Schlich *et al.* (2007) concluded that leisure travel is becoming more irregular and somewhat idiosyncratic compared with other types of travel. Moreover, they recognised five key findings, namely the: importance of social interactions, the widening of leisure activities impacting on travel, the trend of complexity, variety/novelty-seeking and routines, the suburbanisation and full 'motorization of society' (p. 235). To these we would add the influence of social media and consumer-generated content (Yoo and Gretzel, 2011).

If we unpack each of these ideas, they help provide a more coherent picture of contemporary leisure travel. The first of these, the notion that leisure travel lies within the intra-personal perspective along with the impact of different lifestyles is a key finding. More specifically Schlich *et al.* (2007) identified the importance of social interactions with friends and relatives, including the growing spread of social networks. It is, therefore, the travel decisions made around these cordial networks which appear to be crucial to an understanding of leisure travel behaviour (Kowald *et al.*, 2010). Others have used the social network approach to focus on the individual traveller's propensity to engage in various leisure activities (Carrasco and Miller, 2009). These studies have examined the size, composition and structure of social networks, showing in summary that their composition are clearly related to social–leisure travel. The implication of these studies is that leisure travel cannot be understood adequately simply by linking people's socio-demographic background or indeed the characteristics of the built environment within which they are situated (Ettema and Schwaren, 2012).

A second key characteristic is the trend towards the increasing complexity of leisure travel and activities during the last 20–30 years in most developed economies. The range of leisure trips include: sporting events (both watching and participating), visiting friends and relatives, a range of entertainments, eating out, leisure shopping and daytrips. All these forms of leisure activities accounted for almost 40 per cent of trips in England in (Department for Transport, 2016b). The same survey data shows an increase of 6 per cent in leisure trips in period between 1997 and 2014. Table 6.6 shows the features of certain types of leisure trips in terms of average characteristics by age groupings. As can be seen, holiday and day trips peak for the 60–69 age group reflecting the increased leisure time for many retired and early retired people. Indeed, as the UK's population has growing proportions of older, retired people the propensity for more leisure and holiday trips will increase. The underlying influences revealed by survey data from England show relatively little variation between age and the number of leisure trips, although there is a peak in the earlier years of retirement (ages 60–69), and similarly there are no significant differences between men and women (Department for Transport, 2016b). Authors such as Acker *et al.* (2016) have highlighted the importance of lifestyle on travel behaviour, while Schlich *et al.* (2007) consider the aspect of self-realisation as a key indicator. This is the idea of fulfilment and in this case using leisure trips and activities to fulfil personal ambitions. As to be expected there are strong relationships between higher income levels and more leisure trips. Interestingly, people living in rural areas make more leisure trips, usually because facilities tend to be further away but this does not fully explain why they make more trips than their urban counterparts (Department for Transport, 2016b).

A further important factor concerns high levels of daily variability of leisure trips. This is in part due to many such trips becoming habitual, for example trips

Table 6.6. Average leisure trips per person in England (2015) by age

Purpose	17–20	21–29	30–39	40–49	50–59	60–69	70+
Visiting friends and relatives[1]	150	128	123	112	135	168	130
Sport/Entertainment	63	53	57	56	54	79	59
Holidays/Day trips	24	23	36	41	46	56	44
Total	237	204	216	209	235	303	233

[1]Note: there is no individual data on leisure shopping so shopping has been excluded here.

Source: Department for Transport (2016b) (based on Table NTS 0611).

to leisure and fitness centres. More significant is the linking and timing of leisure activities on a daily basis becoming increasingly complex.

This links to a fourth factor relating to the increasing search in daily life for new experiences, which in turn are prompted by the variety of information provided by social media and mobile technology. For many people the smartphone is integrated into everyday behaviour, including leisure and tourism trips. It helps maintain social activities with families and friends. More especially, MacKay and Vogt (2012) found this technology spilled over from everyday activities into the travel context leading to the complexities of trip patterns and a blurring of spatial boundaries.

The final factor derived from the research of Schlick *et al.* (2004) is the impact on trip patterns of increasing suburbanisation and the full motorisation of society as we discussed earlier in the chapter. This has in large part produced diffuse and less predictable leisure-travel patterns.

Slow tourism and travel behaviour

One of the reactions to many of the leisure and tourism travel behaviours we have examined so far in the chapter has been the development of the concept of slow tourism. This, according to some commentators, 'signifies anti-consumerist displeasures associated with unsustainable lifestyles and eco-desires for different kinds of identities' (Fullager *et al.*, 2012, p. 4; following Scholor (2010) and Soper (2008)). Indeed, the notion of slow tourism needs to be framed within a broader perspective of 'slow movement' within many industrialised societies. It may be viewed in part as a reaction to the ever-accelerating pattern of lifestyles in search of alternative leisure and travel opportunities (Oh *et al.*, 2016).

The origins of slow tourism are in the slow-food and slow-city movements dating from the 1980s initially in Italy following Sawday's book *Go Slow Italy* (2009). This was largely a reaction against the growing impact of fast-food outlets such as McDonalds. Slow leisure and tourism therefore are part of this wider movement emphasising the revolution of leisure time both in terms of an experience and also with a view to being more environmentally sustainable. However, as Fullager *et al.* (2012) point out it would be somewhat naïve to see slow tourism as the antidote to aspects of global travel. Nevertheless, as various studies have shown it is becoming a 'lifestyle' choice for some travellers. There are a number of aspects we need to understand in this context, including understanding the dimensions of slow tourism or more accurately in terms of this chapter slow travel. The attempts to define and increase our understanding of slow tourism are varied. Lumsdon and McGrath (2011) for example, argue that its concept is stronger among 'writers, small-scale entrepreneurs and growing numbers of independent tourists' (2011, p. 266). In terms of attempts to define the underpinning concepts of slow tourism we can recognise a number of attempts, but as Table 6.7 shows although there are differences there are significant commonalities.

Table 6.7. Principles and characteristics of slow tourism

• Taking fewer trips
• Development of low-carbon travel
• Enrichment of the travel experience (Principles according to Dickinson and Lumsdon, 2010)
• Experiencing the journey as part of the holiday
• Valuing and learning local culture
• Slowing down and taking time to relax
• Minimising impacts on local communities and the environment (according to Lumsdon and McGrath, 2011)
• Slow travel is a state of mind
• Tourists should travel slow and avoid flying
• The journey is intrinsic to the tourist experience
• Locality is important
• Slowing down to enjoy the city or the landscape is a key element
• Culture through language and engagement with locals creates a better holiday
• Tourists should make opportunities to seek out the unexpected
• Giving back to local communities is integral

Source: Modified from Gardner (2009).

One of the main issues is the interest shown in the supply or provision of slow tourism holidays by the supply sector as Lumsdon and McGrath (2011) point out. Their survey of existing trends show slow tourism is a growing market segment. Within the UK Mintel (2009) examined a panel of 1,665 respondents in the UK who preferred overland travel than flying (quoted in Lumsdon and McGrath, 2011). Of course this in itself is a rather weak and imprecise indicator. Other earlier reports expected slow travel would grow by an estimated 10 per cent per annum between 2007 and 2012 (Euromonitor International, 2007). There is no lack of marketing on slow tourism as evidenced by the number of websites, the establishment of the International Slow Tourism movement, and the increasing number of funded projects on slow tourism by the EU (see Dickinson *et al.*, 2011 for a list of related slow-tourism websites).

Empirical studies of the motivations for slow tourism have highlighted some key findings which support the ideas of it being more than a niche market. According to the findings of a study of four destinations in the USA based on 1,068 responses undertaken by Oh *et al.* (2016), motives for slow tourism are driven by 'a mental psychological and behavioural process embedded in a vast majority of goal-driven consumption activities' (2016, p. 215). They go on to argue that motivations for slow-tourism activities have much in common with those linked to general tourism. In contrast a more limited qualitative study in the UK by Lumsdon and McGrath (2011) highlights many differing consumer experiences associated with slow tourism as identified in Table 6.4.

More important to our concerns in this chapter is the study by Dickinson *et al.* (2011) which focused on exploring environmental discoveries relating to slow travel. They also identified so-called hard and soft slow travel based on types of travel choices. In this context 'hard slow travellers' were those that 'made a conscious decision to avoid flying or car use unless absolutely necessary' (2011, p. 287).

'Soft slow travellers' in contrast generally behaved in an environmentally sustainable fashion but tended to embrace a number of types of holiday travel, such as flying and by car. The more general findings are that, first, the mix of slow-travel components (see Table 6.3) vary depending on the individuals and the leisure and holiday context. Second the key component depends on the destination and more significantly on the importance of the travel experience. Third, such experiences relate to the enjoyment of the journey itself and the levels of engagement with communities during the journey and at the final destination. Finally, environmental concerns are a key component and therefore a key marketing aspect for promoting slow travel. Given the growth of slow tourism it is not surprising that new or older reinvented forms of mobilities have emerged. For example, O'Regan (2011) has noted the resurgence of hitch-hiking as a form of slow tourism practice originating in the 1920s. However, while this involves walking it also makes use of getting free rides in cars. But it should also be noted that organised walking holidays have also increased their importance during the last decade. However, although walking holidays have become part of the slow-tourism agenda, walking as a general activity has declined in countries such as the UK over the last 30 years mainly as car ownership has increased.

In contrast to this general decline, walking as a leisure activity has become more popular, with the proportion of adults walking at least five times per week for utility and recreation increasing by 6.9 per cent between 2012/3–2015/6 (Department for Transport, 2016c). It is difficult to establish the reasons for this increase on the basis of these survey results. Indeed, there are mixed messages from the National Transport Survey on decreases in walking trips while the Active People Survey (APS) shows an increase in frequent walking. Such difference relates to data capture and 'how walking trips are defined and measured' (Department for Transport, 2016b, p. 6).

The other main form of slow travel is cycling and there has been a growing interest in cycling in the UK due, for example, to the influence of successes in sport. The British winners of the Tour de France along with the Olympic successes in cycling have captured the imagination of younger people. Similar trends have been noted in Australia where 10.5 per cent of the population participate in cycling (Faulk *et al.*, 2006). More significantly there have been recent attempts to develop cycle tourism. Across Europe according to Eijelaar *et al.* (2010), cycling is increasing in popularity due partly to more regional and national cycle networks and routes. Their work is based on research conducted across Europe making a distinction between cycle holidays (including overnight trips) and day trips. They identified the main outbound markets as the UK and Germany, with Austria, Denmark and France being the predominant receiving destinations. In this study although the concept of slow tourism is not discussed directly they recognise the contribution to sustainability along with its developmental impact on rural tourism. One telling statistic is that a German cycle tourist will 'produce 66% less holiday emissions [CO^2] than the average German holiday maker' (Eijelaar *et al.*, 2010, p. 15). Furthermore, the carbon footprint relating to domestic cycling holidays in the Netherlands is 35 per cent less than that for the average domestic holiday (Eijelaar *et al.*, 2010).

Slow tourism is viewed at one level as a reaction to 'fast' tourism by attempting to slow travel down and encourage more 'reasonable' (Banister, 2008) travel times by using alternatives such as rail travel for holidays. Indeed, according to Dickinson *et al.* (2011, p. 294), 'the ingredients of slow tourism will vary depending on individuals and context'. For example, for some slow travellers environmental issues may be a key determinant that divides the mode of travel and the need to reduce their carbon footprint therefore determines the modal choice. However, this is also coupled by other important motives such as the desire to enjoy the journey to the holiday destination.

Conclusion

The impact of changing travel patterns for both tourism and leisure trips has been profound both in terms of growing complexities of behaviour and more importantly on the environment. The growing addiction to air travel coupled with its relative cheapness as part of holiday costs has done much to increase such impacts, similarly the notion of 'auto' mobility as outlined by researchers such as Urry (2007) has provided further impacts on such environmental impacts. Travel in tourism and leisure is part of post-modern lifestyles across a range of developed and emerging economies such as China and India. Indeed, the notion of tourism inducing different patterns of travel behaviour has been noted within the chapter, especially in terms of those individuals who do not appear to take their sustainable behaviour patterns from home to holidays. Similarly, the chapter has noted how leisure trips have increased and become more complex.

Discussions within the mobilities literature have done much to aid our understanding of the complexity of their various travel patterns, particularly within the context of the car. Added to this is the influence of mobile technology on travel, and as Jansson (2007) argued, such technology changes the nature of travel by 'decapsulating' the experience of tourists, since the feelings of adventure and escape are opened up. This happens in two ways: first by providing information to the tourist and second by allowing the tourist to communicate their experiences via Facebook, Instagram or Twitter to their friends and relatives. Indeed, the research by Wu and Pearce (2016) on the international travel patterns of Chinese tourists has shown the growth of a more expansive second wave of Chinese tourists aided by mobile technology.

In contrast to these widespread trends are the niche but growing ideas about slow tourism. Authors such as Dickinson *et al.* (2011) and Becken (2007) argue that reducing the carbon footprint of tourism will not be easy. Indeed, as Dickinson *et al.* (2011) point out from their research: 'Slow travellers ... exhibited a keen ability to manoeuvre arguments to suit their travel practice and maintain a credible self-image' (2011, p. 295). In considering how we might pursue more sustainable forms of tourism, it is clear that tourist identities are key to framing choices and experiences, many of which still rely on high levels of consumption and the central importance of self-image.

Part III
Sustainable mobilities

7 Sustainable mobility

The policy challenge

Introduction

This final part of the book provides a critical analysis of how our understandings of contemporary transport geography and mobility, as evidenced in Parts I and II, can inform our understandings of contemporary and forthcoming debates on reshaping our society for a lower-carbon future. In this way, this and the subsequent two chapters pose critical questions about how contemporary understandings of mobility can change within the context of existing infrastructures, technologies and social relations. In so doing, these chapters highlight why studying transport and mobility is so central to the social transformations that would seem inevitable to deal with global climate change and energy scarcity.

This chapter provides a contemporary transport planning policy context for what we have come to know as sustainable mobility, again focusing on the UK. It begins with a background of the role of individuals in responding to policies relating to sustainable mobility. Then the chapter delves into transport-planning policy development. Aspects of contemporary policy instruments are considered before the concluding section. These include Local Transport Plans, the role of target-setting, and the process of development control.

Linking the individual to sustainable mobility policy

As demonstrated earlier in the book, the rapid rise in car-based travel in the twentieth century stems from individual travel demand for mobility. Behavioural change is an integral concept within contemporary mobility policy, which encourages or forces individuals to change towards more sustainable forms of transport. This is explored further in the next chapter. However, it is not that easy to get individuals to change travel behaviour when there are some underlying aspects to consider. Primarily, there is an implicit aspiration that individuals will make more sustainable travel choices, but this is not necessarily the case, particularly when most individuals are not altruistic in their decision-making, behaving according to self-interest rather than for society as a whole. For instance, in response to implementation of a road-user charge (e.g. for sustainable mobility benefits), individuals may make longer journeys by a local road rather than a trunk road with the charge and actually add to the levels of carbon emissions and rural congestion. Ultimately, dependency on car-based mobility is very difficult to break, and few individuals would change travel behaviour unless there is an obvious benefit to them, typically in terms of cost and/or time. Furthermore, many individuals are constrained by having no alternative to private vehicle transport for certain trips. It could be that where they live or work, or that their job, means that they have to use motorised transport.

If there is a background trend, as there has been over the last 30 or so years, of people changing behaviour from the sustainable modes of walking, cycling and public transport to the motor car, then it is very difficult to reverse this pattern through the promotion of sustainable mobility. Mobility policy may also witness individuals shifting between sustainable transport modes, which need to be accounted for in policy instruments. For example, cycle use may increase, but only at the expense of bus travel. Another difficulty associated with sustainable mobility policy is that outside factors have a large impact on behaviour, in a positive or negative manner. One recent impact upon travel has been the economic recession following the Global Financial Crisis of 2008 which has affected the way in which individuals travel.

When it comes to theoretical and modelling approaches (examined further in the next chapter), there are a couple of underlying individual-based difficulties. First, individuals are irrational, which means they might not make the same choice, even when given identical trip information. This goes against a core assumption of most transport-related choice models, which is that individuals act rationally. Second, people tend not to be static in their decision-making, and change in the ways they respond according to a new set of circumstances (e.g. implementation of a road-user charge) in different ways than previously.

An environmental segmentation study by the UK Government (Department for Environment, Food and Rural Affairs, 2008) demonstrates responses when sustainable mobility options are tested against other sustainable behaviour measures. The 12 actions examined cover improvements to household efficiency, reductions in the impact of food consumption, and reductions in the emissions from travel. Three of the actions concerned sustainable mobility: use more efficient vehicles, use car less for short trips, and avoid unnecessary flights (short-haul). When the 12 actions were plotted on a graph of actual behaviour versus carbon-emissions impact, the three transport mobility actions were the only ones in the quadrant with low levels of behaviour against high carbon-emissions impact. This demonstrates that sustainable mobility policy instruments, across other sectors, tend to start with the lowest base of behavioural level and have the greatest negative environmental impact.

Empirical evidence shows that most of the population in the UK are unwilling to change travel behaviour. It has been shown (Davison and Ryley, 2010) that only a small segment of the population are trying to fly less for environmental reasons (8 per cent in their survey of residents from the East Midlands region). Furthermore, Barr *et al.* (2010) demonstrated that even the most committed environmentalists who behave sustainably in other areas of their life such as with recycling, are unwilling to stop their low-cost air travel.

Groups of the population can be considered from a social perspective too, and this can be translated into transport-policy instruments. There has been a growing recognition that transport can be a significant barrier to social inclusion. Around the time of the integrated transport strategy, a UK government body, the Social Exclusion Unit, was established (in 2001) to explore, and make recommendations to overcome, the problems experienced by people facing social exclusion in reaching work and key services. The Social Exclusion Unit (2003) examined the links between social exclusion, transport and the location of services, mainly concerned with the accessibility of local services and activities. It particularly focused on access to opportunities with most impact upon life-chances, such as work, learning and healthcare.

Associated with social need has been accessibility planning policy, the ability of individuals to reach a range of services and facilities easily. The core spatial element of accessibility analysis, make it a key technique within transport geography. The need for accessibility planning stems from the fact that for most people, at most times of day, for most trip purposes, accessibility has been getting worse in recent years (Halden and Davison, 2005), clearly linked to the increase in car-based mobility. Accessibility planning is the process by which this trend is reversed, securing improvements for all people but particularly for those in greatest need (i.e. the socially excluded). Accessibility is an important aspect of transport planning policy (e.g. planning the accessibility of new developments) and is a mandatory aspect of Local Transport Plans.

Development of transport planning policy in the UK

The 1947 Town and Country Planning Act nationalised development rights (Headicar, 2009, pp. 77–78). To some people this was a negative step, by preventing and limiting development and imposing stricter conditions. With respect to transport, the increase in travel, particularly by the motor car, led to the demand for new developments. The problems of land use and transport leads to development control issues, greater in large towns and urban areas. This conflict is most visible on the urban fringe where developers want to be, near the motorways and airports, but this is also where greenbelts lie. The tension of whether to protect or build upon greenbelt land is particularly pertinent given the shortage in (affordable) housing and the lack of available brownfield land. The UK (unlike for example the USA) has a high population density and so suffers from a lack of available land for development. Pressures are greatest in London and the south-east of England, and one pertinent transport example is the ongoing debate surrounding where to further develop airports given the lack of space. A new runway at existing airports such as Heathrow or Gatwick require solutions such as tunnels for surrounding motorways (to overlay the runway), while the other option is the expense of building a brand new airport on new land such as an island in the Thames Estuary area.

The geography of urban areas is important, in terms of where to locate developments, for instance in the centre or in the suburbs. The more sustainable mobility modes of walking, cycling and public transport are more suited to developments in the centre of towns and cities. Cities designed in compact form promote a more sustainable mobility, in association with promotion of the city centre, the construction of high-density buildings and imposition of a surrounding greenbelt. A further, more recent, step is to develop a 'car-free' city or at least restricted motor car access and parking. Again, the development for sustainable mobility goes against underlying land-use trends such as an increase in out-of-town shopping centres and lower-density housing. It is particularly prominent for supermarket location, given that the weekly food shop is the most car-dominant trip purpose.

In response to the increasing spatial challenges, the UK central government produced a series of PPG (Planning Policy Guidance) notes during the early 1990s. The note relating to transport, PPG13, was introduced in 1994. It clearly brought sustainable mobility to the fore in transport planning, with three basic aims: to reduce growth in the length and number of motorised journeys; to encourage alternative means of travel that have less environmental impact; and to reduce reliance on the private car (Department for Transport, 2011). Subsequent versions (published in 2001 and 2011) did push more car restraint, particularly with maximum parking standards for new developments, but must still be considered as guidance rather than requirements for developers. This is a repeated problem for planning policy relating to sustainable mobility – it has tried to *encourage* change, when *forcing* it would have been more effective.

One change over the last 20 or so years has been the increase in the role of the private sector in transport-planning policy, which has different goals from the public sector bodies involved, such as local authorities and central government (e.g. the Department for Transport). The public sector is interested in wider social, economic and environmental objectives, in contrast to the private sector, which will naturally have a primary concern with maximising profit. The private sector has various forms in transport planning, not only transport companies such as train and bus operators, but also other stakeholders such as local businesses and developers.

Another change in transport-planning policy has been an increase in public participation. For instance, for local traffic-calming schemes, a local authority must conduct information gathering and research exercises, involving the public using questionnaires, interviews and group discussions, say through residents' liaison groups or consultative committees, and public meetings and exhibitions (Booth and Richardson, 2001). The impact varied geographically around the UK. From a survey of Local Authorities, although authorities were quick to involve the public (Bickerstaff *et al.*, 2002), it was hard to identify the roles of the people involved, whether as partner, stakeholder or member of the public and transport planning schemes can still favour a vocal unrepresentative minority. Hull (2005) assessed the transport-planning aspects of the urban areas of Bristol and Newcastle, and how they translate from concept to reality. In these areas, it was considered that too much was localised (i.e. not conurbation or city-region based) and at the time regional institutions were still 'bedding down'. Two underlying themes emerged: the fact that there was no legal requirement for sustainable development, and the way that transport planning policies were too often affected by political changes through local elections. A final comment on public participation is that even though there has been an increasing role of the public in planning, much of it has concerned NIMBY-ism (Not In My Back Yard) whereby people living near to a planned transport project campaign loudly for it to be implemented elsewhere.

It can be clearly stated that the spatial, geographical element is central to transport planning, and cuts across international, national, regional and local levels. The international dimension for the climate-change phenomenon has been discussed in Chapter 3. For a national transport-planning policy context there are four spatial levels within the England and Wales system (a similar situation existing in other areas within the UK, but under different policy guidelines): National, Regional, County and District. Figure 7.1, using the town of Loughborough as an example, demonstrates these examples. The top two levels (National and Regional) relate to guidance. The role of central government is to make policy to guide local authority planning through PPG and then more recently the National Planning Policy Framework (Department for Communities and Local Government (DCLG), 2012). Central government also has the power to 'call-in' local authorities in relation to their plans or planning decisions in order to have them modified or reversed. The UK planning system could therefore be permissive rather than prescriptive, a system of development plans that guide development. This means that planning permission is not necessarily given if a proposed development conforms to the relevant development plan. There could be a situation of a supermarket putting similar superstore plans in two separate towns, which have similar development plans, and yet only one of them is granted permission to develop.

At the regional level, voluntary, multi-party and inclusive Regional Chambers were established in 1998 in each of the eight English regions outside London, but were then abolished in March 2010. Therefore, for some period it was considered that some of the planning power could be devolved from central government to the regions to better cover some of the transport issues among other planning elements. However, with the extra bureaucracy involved in the regional level,

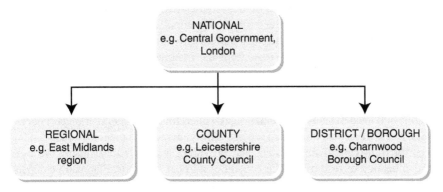

Figure 7.1. Planning levels in England and Wales, using the example of the town of Loughborough.

associated with financial pressures from the Global Financial Crisis (from 2008 onwards) and political imperatives to streamline procedures, this planning level has now largely been removed from the structure.

The role of Local Transport Plans

This section explores how national transport-planning policy is implemented through the development of Local Transport Plans at a more localised level. Such plans are intended to set the aspirations of local areas for their transport needs and as such form the basis of the interaction between the planning system and the provision of transport infrastructure locally. A Local Transport Plan is typically implemented at the county level (e.g. Leicestershire).

The 1998 Integrated Transport White Paper marked a shift in transport policy, and included requirements for UK local authorities to produce five-year Local Transport Plan strategies, and implemented through the Local Transport Act 2000 (Marsden *et al.*, 2012). The previous system, Transport Policies and Programmes, was an annual process and it was thought to be more efficient to have a five-year Local Transport Plan process (Headicar, 2009). That stated the Local Transport Plan does have an annual progress reporting system, albeit as a much lighter touch than previously. The five-year planning process does provide a greater certainty of future funding for authorities.

Guidance on the development of Local Transport Plans (Department for Transport, 2004b) emphasises four key themes: setting transport in a wider context; identifying the best value for money solutions; devising locally relevant targets; and developing indicators and trajectories. For the last theme, as with the sustainable development indicators (see Chapter 3), there are a series of questions about devising a fair and appropriate list of relevant indicators to measure transport-related performance. Mandatory indicators for all authorities were an accessibility target, a change in area-wide road traffic mileage, cycling trips, mode share for the journey to school, and bus punctuality. Other indicators include peak traffic, congestion and air quality. Over time, the central government have been developing the evidence base and rigour of measurement within Local Transport Plan data collection, to ensure an improved set of indicators.

Local Transport Plans naturally reflect the transport policy development towards sustainable mobility (e.g. from the Integrated Transport White Paper), so they emphasise integrated transport solutions that encourage public transport, cycling and walking. They also reflect the increasingly inclusive approach within transport planning policy, as discussed earlier in the chapter, by having greater public and local business participation as well as involving public transport operators. Local Transport Plans also represent the general movement towards relaxing control from central government and providing a greater local discretion over the allocation of resources. The content of a Local Transport Plan can be dependent on a key major scheme, which is fine when implemented, such as the Light Rail scheme in Nottingham (within Nottinghamshire Local Transport Plan), but they can look out-of-date very quickly if a scheme is rejected, such as congestion charging in Edinburgh (within their Local Transport Strategy, the Scottish equivalent of a Local Transport Plan).

Over time, central government (through the Department for Transport) adjusted Local Transport Plan funding allocations to local authorities by plus or minus 25 per cent based on the quality of the plans (Marsden *et al.*, 2009). This study also reported that local authorities considered that too much was being expected of them by the Department for Transport. In essence, local authorities are competing with each other, but as they are very different in nature from each other the criteria become very subjective. For example, the Local Transport Plan for the rural county of Rutland has a very different content from the document for Nottinghamshire, which contains the large city of Nottingham and associated transport issues such as congestion. The Local Transport Plan therefore effectively becomes both a bidding document to central government and an overview of strategic planning for the local audience.

The original Local Transport Plan five-year cycle means that currently local authorities are now on their third Local Transport Plan, from 2011. As stated by Simm (2010) they have a longer time horizon but with finances updated annually. It should be noted that there has been a major reduction in funding for Local Authorities from central government due to the recession that followed the Global Financial Crisis. Local Transport Plans now consist of separate but inter-linked documents, with a long-term strategy to 2026, but supported by rolling three-year implementation plans (including annual capital spending).

Target-setting in transport

Target-setting is a relatively recent phenomenon in transport planning policy, over the last 15 years or so, such as through Local Transport Plans where they constitute one of the key aspects. In the transport academic literature, several papers review targets (e.g. Humphreys *et al.*, 2005; Marsden and Bonsall, 2006; Marsden *et al.*, 2009), but there has not been a rigorous approach to target-setting in practice.

Setting targets could be considered more of political game-playing than a precise science. Targets may be set under certain circumstances, but whether these targets are to be achieved is a separate issue, particularly when the targets are not legally enforceable. When considering transport targets, the 1996 National Cycling Strategy published by the Department of Transport springs to mind (Department for Transport, 1996). The strategy included targets to double cycling mileage by the year 2002 and quadruple it by the year 2012 from 1996 UK cycle-trip figures (average of 16 bicycle journeys per person per year). These targets were subsequently scrapped with a lack of commitment from the UK government. Therefore, targets need to be achievable and realistic at the outset.

When setting targets they can be set as an aspiration not realistically attainable, or as an achievable aim with little encouragement. Hopefully those within Local Transport Plans can fall somewhere between these two levels, and realistic to the local situation. Also, there is a difference between actual or percentage increases, and even within a percentage change it could be 'increase by X per cent' or 'reach X per cent of modal share'. A large percentage increase may seem a great improvement, but not if the starting point is a small baseline.

Targets are useful for politicians to show that there has been success in a particular measure, but the level at which they are set is important. An example can be shown from the Leicestershire Local Transport Plan (Leicestershire County Council, 2015) and Key Performance Indicator 4: Reduce total casualties on our roads by 29 per cent by 2020 (from a 2005–2009 baseline). The Local Transport Plan could have contained a much harder target of say 35 per cent. Think about the following two scenarios for traffic congestion reduction:

1. A target of 29 per cent casualty reduction, then achieve a 30 per cent reduction in 2020. The politician is happy that the council has met the target.
2. A target of 35 per cent casualty reduction, then achieve a 32 per cent reduction in 2020. The situation is better because there has been a greater reduction (perhaps the council have worked harder to achieve the target), but the politician is not as happy because the target has not been met.

The second scenario has a preferable result, but the first is the one that a politician would desire. This demonstrates the importance of politics within UK transport-planning policy.

Conclusion: contemporary transport planning policy-making for sustainable mobility

The chapter has covered transport-planning policy and the increasing importance of sustainable mobility. The background context of the role of individuals in responding to policies relating to sustainable mobility was initially provided, which is developed further in the next chapter relating to deeper aspects of behavioural change. Contemporary transport policy-making for sustainable mobility has a challenge stemming from a need to better understand individual response to such policies. Many sustainable mobility initiatives have struggled over the last 20 or so years in the UK because of underlying attitudes and behaviour towards travel that have been hard to shift. This is a repeated problem for sustainable mobility through transport planning policy guidance (e.g. PPG) – it has tried to encourage change, when forcing it would have been more effective. In terms of wider sustainable development, two underlying themes emerge: the fact that there is no legal requirement for sustainable development, and the way that transport-planning policies are too often affected by political changes through local elections.

Geography has been at the heart of many of the contemporary issues within transport-planning policy. There is the underlying spatial structure of the planning system from the local to the national (and to climate change at the global level), as well as the spatial variations that exist between local areas, as demonstrated between the different emphasis within neighbouring rural and urban Local

Transport Plans. Other spatial pressures have been identified within this chapter, including the high levels of population density in London and the south-east of England (and associated transport problems such as determining where to develop airports) and the tensions that exist with sustainable policy for urban areas with car-based mobility encouraging sprawl (further examined in Chapter 9). Finally, accessibility planning has ensured geographical analysis is a key component within transport-planning policy that encourages sustainable mobility.

8   Sustainable mobility

The challenge of behavioural change

Introduction

The challenge of promoting sustainable mobility is one that has emerged from the incorporation of sustainability principles into the workings of central government and as such is reflective of the political strategies that have been developed to make sustainable development a workable political project (Connelly *et al.*, 2012). As this chapter, along with Chapter 9 will outline, we can identify three broad approaches that have been taken towards the promotion of sustainable mobility. These are: the development of behavioural change programmes for individual citizens; the alteration and reshaping of the built environment to promote sustainable mobility; and the use of smart technology to reduce carbon emissions from transport. We will focus on the first two of these approaches in this book because we argue that the challenge for society is its obsession with hyper-mobility and the ways in which places have been planned that promote increasing mobility. We argue that these two processes are unlikely to be suppressed or conceptually challenged through the promotion of 'business as usual' approaches offered by smarter technology, which often fails to tackle the embedded nature of mobility as a social practice (see Chapter 1). Accordingly, we argue that understanding the dynamics of behavioural change and the associations it has with the built environment is key for radically changing our mobility futures.

In this chapter, we aim to explore the first of these approaches, relating to the seemingly ubiquitous topic of behavioural change. In so doing, we first examine the historical development of behavioural change as a political strategy, linking it to developments in governance arrangements. We then examine two contrasting approaches to understanding behaviour that have been developed in the social sciences, related to social psychological and sociological framings of behaviour or practice. In each case we examine how transport and mobility researchers have utilised these approaches and the intellectual challenges that each raise, alongside the problems associated with practically implementing change using such approaches. The chapter ends by providing a discussion on the ways in which social scientists might come to use methods of co-production to collaboratively produce more effective strategies for change in a more politically engaged manner.

Doing your bit?

Perhaps the first question we ought to pose in this chapter is why we are even discussing the idea of behavioural change? If this text had been authored 20 years ago, changing behaviours might have merited a mere footnote as a strategy that could have some possible potential. Yet it now seems as if policy-makers at all levels of government and in non-governmental organisations have wholeheartedly embraced behavioural change as a strategy for approaching the ambitious carbon-reductions targets they are tasked with meeting (Jones *et al.*, 2013). To understand the development of behavioural change as a strategy, we need to look back almost 30 years to developments that occurred in countries such as the UK and USA, which have led to the overt focus on the individual as a reference point for ecological salvation. This is without doubt an exercise in appreciating how behavioural change is an implicitly political project, one that connects with contemporary modes of neo-liberal governance and individualisation (Whitehead *et al.*, 2011).

As Giddens (1991) noted over 25 years ago, the 1980s witnessed a fundamental shift in the ways central government viewed its relationship with citizens in nations such as the USA and UK. In what has now been widely referred to as a shift towards neo-liberal government (Rose and Miller, 1992), the newly elected right-wing conservative administrations in both the UK (led by Margaret Thatcher) and the USA (led by Ronald Reagan) championed a return to what they argued were traditional values of small government. This involved the idea that the nation state should return to a naturally smaller state, enabling individuals to exercise the freedoms that were their right as citizens in a liberal democracy. This necessarily required shrinking the role of the state and an increasing role for private capital, enterprise and individualism (Jessop, 2002). Accordingly, in the UK a programme of mass privatisation of state-owned industries (e.g. British Airways, the National Bus Company and ultimately British Rail) was initiated that fundamentally altered the balance of power. In what Jessop (2002) has referred to as the 'rolling back' of the state, individual citizens were given the right to have a stake in what were once state monopolies. Indeed, in popular culture, the 1980s and early 1990s were a time of considerable enrichment for what became known as 'yuppies', immortalised in the British comedian Harry Enfield's 'Loadsamoney' sketch (Youtube, 2008b).

For a short period, the emergence of a rampant individualism seemed to be an effective tool for economic growth, but as Etzioni (1993, 1995) pointed out after one of the worst British economic recessions for several decades, it would not be possible for a shrunken state to achieve the necessary policy goals on everything from healthcare and social welfare to the environment if individuals were solely concerned with their own wealth accumulation. This led to the development of the notion of the citizen-consumer (Clarke *et al.*, 2007), a tool first utilised by government by the then British Conservative Prime Minister John Major and continued by Tony Blair, leader of Labour's administration in the UK after 1997.

The citizen-consumer concept built on Etzioni's (1993, 1995) idea of communitarianism, in which the rights of free individual consumers were set alongside responsibilities for and to others. As early scholars of the concept have noted (Scammell, 2000; Slocum, 2004), the citizen-consumer is built on the neo-liberal assumption that individual choice is the cornerstone of a free society and that it is not the role of the state to determine courses of action through regulation or financial penalties. Yet it is also important for the state to meet certain policy goals, everything from carbon emissions reductions to obesity reduction. In this way, it must develop strategies for guiding people in their choices and the way this has been framed in countries like the UK and the USA is through citizenship.

The compelling logic of the citizen-consumer is that if real change is to be effected, it has to be undertaken not at the level of collective consciousness, but by appeals to the very individuals that the neo-liberal state believes to be so important. This logic is largely responsible for the rapid development of behavioural change programmes that have proliferated across government departments and percolated down to local authorities and government agencies. Within the UK, the evocation of choice has been institutionalised through the most recent sustainable development strategy 'We all – governments, businesses, families and communities, the public sector, voluntary and community organisations – need to make different choices if we are to achieve the vision of sustainable development' (DEFRA, 2005, p. 25).

Indeed, building on this signal, both the Department for Environment, Food and Rural Affairs and the Department for Transport have pioneered research that has sought to understand the influences on individual behaviours and to use these for developing behavioural change frameworks (Department for Environment, Food and Rural Affairs, 2008; Department for Transport, 2011). Indeed, for some years the UK's Cabinet Office, with a legacy of interest in behavioural change (Cabinet Office, 2004), hosted the now privatised Behavioural Insights Team, which uses research from the neurosciences and psychology to develop understandings of individual behaviours. Indeed, this unit has received recent impetus from the landmark publication of Thaler and Sunstein's (2008) now well-known book *Nudge*, in which the authors outline the ways in which behavioural change can be initiated through changes in what they refer to as choice architecture. They explain nudge like this:

> … any aspect of the choice architecture that alters people's behaviour in a predictable way without forbidding any options or significantly changing their economic incentives. To count as a mere nudge, the intervention must be easy and cheap to avoid. Nudges are not mandates. Putting the fruit at eye level counts as a nudge. Banning junk food does not.
>
> (Thaler and Sunstein, 2008, p. 8)

Nudge has attracted considerable attention recently (House of Lords Science and Technology Committee, 2011) and is built on the assumption that choice architectures can and should be amended to shift behaviours. In this way, it is a form of what French *et al.* (2009) have referred to as 'social marketing' (Andreasen,

2006) in which knowledge, expertise and insights from conventional marketing are utilised to promote a social good (National Social Marketing Centre, 2016). Along with Nudge, social marketing has had a major impact within the policy community, as evidenced by the ways in which both the Department for Environment, Food and Rural Affairs (2008b) and Department for Transport (2011) in the UK have adopted its principles to develop behavioural change strategies.

Social marketing has three main constituent elements, notably audience segmentation, the development of a good marketing mix and the setting of specific behavioural goals. A useful illustration of this approach is provided in Department for Environment, Food and Rural Affairs's (2008) *Framework for Proenvironmental Behaviours*. This strategy was designed to identify key behaviours that related to carbon emissions reductions (specific behavioural goals) and the segments of the UK population that broadly held similar attitudes and behaviours (segmentation) (Figure 8.1). Using this information, a framework for promoting change was developed based on both current levels of willingness to participate in given behaviours and the impact on carbon emissions.

This framing of behavioural change has become commonplace not only for environmental issues, but across whole sections of government, with health promotion being one of the most well-known examples (Paek *et al.*, 2014) and in recent years, these approaches have begun to receive extensive commentaries and critiques. These have generally taken two forms. In the first instance, there have been critiques of the philosophy underpinning neo-liberal approaches to behavioural change and the assumptions which they make. Among some of the most vocal critiques have been researchers from Aberystwyth University in the UK, who have argued that the UK government's approach to engaging individuals is a form of what they term Libertarian Paternalism (Jones *et al.*, 2011a, 2011b; Whitehead *et al.*, 2011). In this way, they argue that the politics of

Figure 8.1. DEFRA's *Framework for pro-environmental behaviours.*

Source: DEFRA, 2008b. Contains public-sector information licensed under the Open Government Licence v3.0.

choice architecture has reflected the drive in government to create incremental approaches to behavioural change that are focused on specific policy goals. Such goals reflect what Robinson (2004) has termed a passive form of sustainable development, quite different from what the pioneers of sustainability had envisioned in the Brundtland Report (WCED, 1987):

> The concern here is that sustainable development is seen as reformist, but it mostly avoids questions of power, exploitation, even redistribution. The need for more fundamental social and political change is simply ignored. Instead, critics argue, proponents of sustainable development offer an incrementalist agenda that does not challenge any existing entrenched powers or privileges.
>
> (Robinson, 2004, p. 376)

In building on this critique, Slocum (2004, p. 765) argues that current framings of behavioural change reflect '… an outgrowth of classical liberal theory that universalises the logic of the market for all institutions', creating 'passive' citizens and challenges to progressive politics. Indeed, Johnston (2008) argues that such an approach '… implies a social practice that can satisfy competing ideologies of consumerism (an ideal rooted in individual self-interest) and citizenship (an ideal rooted in collective responsibility to a social and ecological commons)' (2008, p. 232).

Accordingly, there has been considerable critique of the evocation of citizen-consumers as agents of change and it is without doubt that the form of Libertarian Paternalism that Jones *et al.* (2011a, 2011b) refer to is important to note, since it reflects and restricts the ways in which we think about behavioural change. However, it is also the case that a second set of critiques has emerged focused directly on the strategies deployed by nudge theorists and social marketing practitioners, which deal directly with the assumptions about the effectiveness, from a pragmatic perspective, of trying to use behavioural change to meet environmental targets.

Barr *et al.*'s (2010, 2011a, 2011b) analysis of social marketing for pro-environmental behaviour change has highlighted some of the pitfalls of relying on forms of audience segmentation through social marketing within the context of sustainable mobility. Their research examined the attitudes and behaviours of so-called 'committed environmentalists' and 'mainstream environmentalists' (Barr and Gilg, 2006), who were individuals reporting high levels of environmental commitment in and around the home in terms of recycling practices, energy and water conservation. Yet when these attitudes and behaviours were compared to those for holiday-taking and travel for holidays, those most committed to environmental action in and around the home tended to be those who flew more frequently for holidays. Such a finding reflects the trend in the UK for those with more disposable incomes to be those who have increased their flying since the advent of low-cost airlines (Graham and Shaw, 2008), but they also tend to be those households who are also more inclined to pro-environmental practices in the home (see discussions in Chapter 6).

Accordingly, as Barr *et al.* (2011a) argue, rigid forms of segmentation through the inflexible adoption of social-marketing principles is highly problematic and negates the importance of different sites of practice; such sites reflect the different consumer environments in which practices emerge, so that the apparent contradiction of environmental concern and taking long-haul flights can be easily reconciled through various psychological processes, such as denial (Stoll-Kleemann *et al.*, 2001) and the contestation of climate-change science (Becken, 2007).

There are also pragmatic questions about the effectiveness of social marketing. Peattie and Peattie (2009) have argued that the major changes required to meet ambitious climate change targets are unlikely to be met through adopting approaches that utilise marketing messages that are conventionally adopted to increase consumption. In this context, they argue that much more ambitious approaches are required if the shifts in consumption required to reduce carbon emissions are to be realised.

Such a view is reflected by some practitioners as well as academic research, as Crompton and Thøgersen (2009) argued in a critical report on behavioural change:

> The comfortable perception that global environmental challenges can be met through marginal lifestyle changes no longer bears scrutiny. The cumulative impact of large numbers of individuals making marginal improvements in their environmental impact will be a marginal collective improvement in environmental impact. Yet we live at a time when we need urgent and ambitious changes.
>
> (2009, p. 6)

These views have encouraged researchers like Dobson (2010) and Seyfang (2010) to advocate a more radical social-change agenda that up-scales the issue of behavioural change to one that focuses on broader economic issues along with the recognition that if significant changes in the way citizens use resources are to be made, it is likely that the underpinning contexts on which consumption is based will need to be tackled. This dilemma, between pragmatic incrementalism and social transformation, is one that is reflected in the ways in which academic researchers have explored the issues of behaviour change. Accordingly, the following two sections of this chapter explore the different ways in which researchers have examined the notion of behavioural change. First we examine the psychology of behaviour change and outline the ways in which psychological theories and models have been applied to understand individual decision-making. Second, we introduce research from sociology that has argued for a reconsideration of 'behaviour' through theories of social practice. In so doing, these two sections aim to provide an over-arching context for appreciating why behavioural change, in academic and practical terms, is such a contested and problematic issue.

The psychology of behaviour change

By far the most extensive literature on behavioural change and its use for the promotion of sustainable mobility is that which has been concerned with utilising a psychological understanding of behaviour. The development of the quantitative social sciences during the 1960s, in which geography was characterised by the spatial turn (Kitchin, 2015b), resulted in approaches in the behavioural sciences that sought to adopt the same or similar assumptions from the natural sciences and to apply these to social phenomenon. Such an epistemology adopted a positivist position, in which the certainties and rationalities of natural systems could be incorporated into understandings of the social world.

As a component of quantitative social science, psychology has developed as an academic discipline along these epistemological lines, in which the principles of objectivity, reliability and verifiability have been incorporated into attempts to understand behavioural responses to particular issues. There have been several ways in which this development has taken place in the context of our key interest on travel. First, travel and transport researchers have been keen to explore the ways in which existing psychological models of behaviour are able to predict behavioural responses according to a series of pre-set determinants (Bamberg *et al.*, 2011). As noted in Chapter 4, two models in particular have dominated the landscape of transport psychology. Fishbein and Ajzen's (1975) 'Theory of reasoned action', latterly to be adapted into the 'Theory of planned behaviour' (Ajzen, 1991), is a very good example of a framework that integrates key psychological constructs such as attitudes, behaviours, intentions and norms, as demonstrated

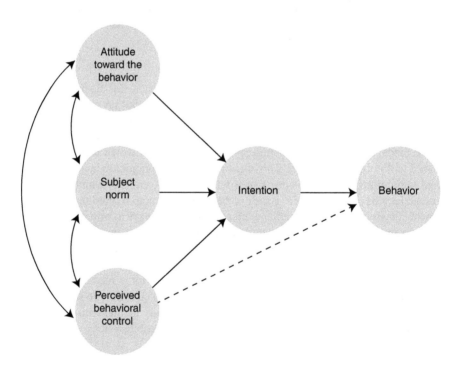

Figure 8.2. 'The theory of planned behavior'.

Source: Ajzen, 1991.

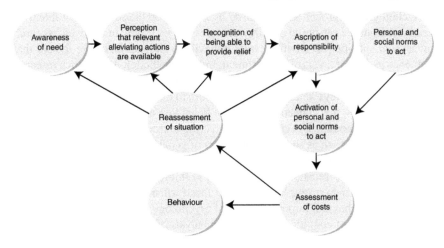

Figure 8.3. A schematic representation of Schwartz's (1977) Norm Activation Model.

in Figure 8.2. Critically, the model highlights the logical positivist nature of this kind of approach, with predetermined directional flows between each construct. Such models are still used widely in transport psychology, for example to predict the adoption of sustainable transport modes (Bamberg *et al.*, 2003; De Groot and Steg, 2007; Heath and Gifford, 2002).

Alongside the 'Theory of planned behaviour', Schwartz's (1977) Norm Activation Model (Figure 8.3) has also been applied to examine pro-sustainability behaviour change, through an examination of the ways in which norms to adopt particular kinds of (pro-social) behaviour are adopted. For example, Nordlund and Garvill (2002) apply Schwartz's theory to understandings of how to reduce personal car use, while Bamberg *et al.* (2011) has highlighted the use of Schwartz's model in a range of transport settings.

A second approach taken in transport psychology has been related to the exploration of a broader set of psychological factors and their influence on behaviour. In this way, authors like Anable (2005), Dallen (2007) and Götz *et al.* (2003) have argued for the integration of a looser set of conceptual tools for understanding travel behaviour that add greater flexibility to the contexts in which they are working. For example, this has enabled the integration of a range of concepts, such as the role of personal and self-efficacy, perceptions of convenience, comfort and reliability, alongside structural and situational factors such as service and information provision.

What these studies have in common is the framing of behaviour and behavioural change around the individual. That is, such research focuses on the cognitive aspects of decision-making that are particular to individual citizens and attempts to appreciate and generalise these. Indeed, much of the research that has adopted a psychological approach tends to do so by focusing on the so-called 'problem' at hand, for example car-based mobility or excessive air travel. In this way, research focuses on specific pro-environmental behaviours and examines the factors that influence particular levels of commitment. Indeed, methodologically such research tends to be quantitative in nature and as such relies on the use of measurements through social surveys and reported behaviours to examine particular constructs.

In recent years, this conceptualisation of behaviour and its deployment through quantitative methods has received sustained critique from researchers in other parts of social science, such as sociology and parts of geography (Blake, 1999; Hobson, 2002; Owens, 2000) drawing on the wider challenges posed by researchers from the field of Science Technology Society (STS) studies (Wynne, 2002). These critiques have focused on four key problems faced when attempting to explore and conceptualise behaviour through the lens of psychology. First, the focus on individuals and the role of cognition drives down analysis of the challenge of issues like mobility to personal decision-making and does not make a connection to the social dimension of mobility practice, in particular how apparently individual behavioural decisions can be understood as deeply embedded practices shared across groups or households. Second, in what Owens (2000) has referred to as the rationalisation of lifestyles, such research often makes assumptions about the processes that govern decision-making and what Shove (2010) refers to as the 'factors' that intercede to shape behaviour. Third, researchers have raised methodological questions concerning the ability to measure travel behaviours within their wider context of social meaning (Urry, 2007). In other words, how much can we know by simply measuring whether someone has opted to drive or to use public transport for travelling to work? We might ask whether there is importance attached to the meaning, purpose and experience of a journey. Finally, there are vexed and important questions over the role of what we constitute as evidence for travel behaviour change. It is without doubt the case that those from a more positivistic approach to social research argue that the quantification of results and the use of particular forms of statistical reliability and replicability are crucial and it is certainly the case that most policy-makers at the national level would demand this kind of science. Yet focusing on the qualitative, lived experiences of travellers can reveal a great deal about how mobility is understood and the social contexts in which it occurs. These critiques, which are formed on epistemological and methodological grounds, have argued for a more engaged view of so-called pro-environmental behaviour, which links to the wider social practices of mobility and underlying social and economic processes. It is to this intellectual perspective we now turn.

The social practice of mobility

As noted in Chapters 4 to 6, the epistemological divide between conventional transport geography and mobilities research rests partly on the social context for travel in contemporary society and it is the latter on which researchers working from a sociological perspective have focused. Accordingly, in arguing for a more critical approach to sustainable mobility as social practice, authors like Barr and

Prillwitz (2014) have highlighted some of the research in sociology that has advocated a focus on what Reckwitz (2002) has termed social practices.

Exploring theories of social practice lie beyond the scope of this book, but as Shove *et al*. (2012) have argued, they are concerned with the development of broad trends that are associated with shared practices that develop over time and space. Accordingly, they are intrinsically associated with behaviours, but there is an important recognition that such behaviours have a social context and heritage. Such an up-scaled analysis recognises the role that time plays in mediating the ways in which people interact with material objectives (such as technologies).

As we noted in Chapter 4, a useful way of characterising the difference between mobilities research and conventional travel behaviour research can be provided by exploring the issue of car use. Within the context of behavioural change from a conventional transport geography perspective, the 'problem' at issue would be the decision of individuals to use their cars for a particular type of activity, say commuting to and from work. The challenge would be to encourage individuals to make different choices to change their decisions and adopt an alternative travel mode. From a social-practice perspective, the issue is not framed as one of individual choice or even individualised decision-making, but is rather focused on the development of commuting as an accepted and sometimes desirable social practice. In this way, a far broader set of issues and questions emerge, including an attempt to understand the temporal development of commuting, its embeddedness within particular work and class cultures and the likely trajectory of such practices as technologies and working lives change.

This critical distinction between individual notions of behaviour and social practices helps to characterise the connectedness of practices to both individuals and infrastructures and technologies. As Verbeek and Mommaas (2008) argue:

> Social practices are conceived as being routine-driven, everyday activities situated in time and space and shared by groups of people as part of their everyday life [...] Social practices form the historically shaped, concrete interaction points between, on the one hand actors, with their lifestyles and routines, and on the other hand, modes of provision with their infrastructures of rules and resources, including norms and values.
>
> (2008, p. 634)

Utilising this perspective, sociologists such as Huddart Kennedy *et al*. (2015), Shove (2003, 2010, 2011) and Warde (2014) have argued that a focus on particular 'problems' like the environment or anthropogenic climate change mask the complexity and embeddedness of mobility practices. In so doing, Shove argues that such practices are rarely shaped by explicit concerns about a given issue, but are comprised of shared understandings about desirable and acceptable practices. Indeed, she also argues (Shove, 2010, 2011) that there are major problems in the way that the dominant behavioural-change agenda has been examined in both the academic social sciences and among policy communities. In her critique of the so-called 'Attitude – Behaviour – Choice' (ABC) model of behaviour change, she argues that researchers and practitioners have become too focused on a limiting epistemological and methodological framework for studying what is really of concern – a social transformation in practices that will result in major shifts in resource consumption.

Shove's (2010) central argument is that current modes of exploring and analysing behaviour change are politically unambitious and lack social context. In line with Dobson (2010) and Seyfang (2006), she advocates a research agenda that questions the narrow and individualised presumptions of behavioural change approaches that focus on ideas of 'small change'. Rather, she argues for an active questioning of the economic-growth assumptions that drive the consumer-based economy that other scholars have argued acts as a barrier for engaged environmental citizenship (Dobson, 2010).

Utilising these ideas, Barr (2015) and Barr and Prillwitz (2014) have advocated a repositioning of the debate on the role of choice architectures to focus on the underlying architectures that frame mobility practices. In research conducted in and around the city of Exeter in south-west England, they highlighted how the individualised issues of personal convenience and comfort associated with private car travel related strongly to the restrictive choice architectures of work–life patterns, inflexible working hours and the assumed long-distance commutes necessary to perform the tasks of everyday life. Indeed, participants in the research argued that many of the localised social networks afforded by former configurations of the economy (such as factory working, work-based social clubs and employee housing) had been replaced by a disparate and spatially segregated network of social and work-related activities. This was also compounded by the sense of fatalism about government transport and planning policies, in which individuals were being exhorted to make different mobility choices, but witnessed continued out-of-town shopping developments that were still using a car-based model of development.

Accordingly, a social-practice perspective can offer much to the debate concerning the potential for what has become known as behavioural change for more sustainable travel (Verbeek and Mommaas, 2008). Yet although this perspective has gained significant ground in recent years (see Huddart Kennedy *et al.*, 2015), it still lacks much of the resonance and empirical acceptance from members of policy and practitioner communities, in large part because it is based on much more intensive and qualitative research analysis of data in specific contexts (Wilson and Chatterton, 2011). As such, it is still the case that many practitioners prefer to rely on what they argue are more robust and reliable quantitative evidence from surveys, although the debate in *Environment and Planning A* (Shove, 2010, 2011; Whitmarsh *et al.*, 2011; Wilson and Chatterton, 2011) suggests that the relevance of social practice theory to policy issues is now beginning to become a more focused agenda.

This division between different elements of social science is problematic because it sends both confusing signals to policy communities and it also prevents academics from pursuing both a critical engagement with these issues and making a sustained and significant impact. As such, we end this section by providing four reflections on the ways in which practice theory could be usefully deployed to situate contemporary mobility challenges faced by policy makers in the twenty-first century (see also Barr, 2015). First, we acknowledge and uphold the important role that economic and infrastructural factors play in shaping what are manifested as individual behaviours in relation to transport and mobility; we appreciate the need to understand the important role that long-held, intricately developed social practices play in shaping everyday life. To this extent, changing behaviour is certainly a long-term issue that demands policy-makers to think beyond the short-term behavioural intervention 'box'. As such, an appreciation of how cultures of mobility have emerged and their likely trajectories is critical. From a practical point of view, this is about long-term thinking and an appreciation that decisions about economic growth, planning and environmental management will all send important signals that will influence mobility practices.

A second reflection therefore lies in appreciating the complexity and meaning that underlies certain kinds of travel behaviour. In considering practice-led approaches alongside psychological research, there is the possibility of uncovering much more about the complexity of behaviour, which could unlock new ways of promoting change. A useful example is the promotion of increased walking and cycling through promoting ideas of shared space in built environments. Social-practice theory could help to explain why there can be opposition and animosity within such spaces, as they bring together practices of walking and cycling that have previously been regarded as separate. This might incorporate ideas of how routines in space become established, the role of aesthetics, and how disruption to space can cause conflict. In other words, we need to use a range of approaches and methods to help us bring meaning to travel behaviours and to appreciate the complexity of apparently simple ideas like 'shared space'.

Third, thinking about practice as well as behaviour enables a much wider conversation about how we move and what we want places to look like; the currently formulated behaviour change agenda is dominated by a very narrow and rigid model of prescribed behaviours that appear to be isolated from everyday life and devoid of social meaning. Indeed, we need to ask how realistic it is to promote public transport use or cycling as individual choices when the infrastructure, built environment and wider social practices all appear to be working against such choices. Thinking about social practice might therefore enable us to overcome the challenge of connecting research on mobility and place, and it could lead us to a more formal recognition of how mobility can be suppressed or promoted through design.

Finally, we agree with Johnston's (2008) critique of the citizen-consumer model of behaviour change, when he argues that restrictive and prescriptive behaviour-change policies have led to as a pacifying of publics. We argue here that behaviour change should be something that is derived through a political process of working out collectively what changes are needed and how these can come about alongside other shifts in economic and infrastructural conditions. This inevitably requires a more politically engaged dialogue that recognises some of the structural challenges of promoting behavioural change. As such, thinking about social practice in the context of behaviour change is about thinking long-term, in-depth and openly about what is possible and desirable. It might also lead us to consider whether 'behaviour change' is an adequate or appropriate term to use in describing the challenge we face.

Conclusion: what role for behavioural change?

Perhaps not surprisingly, we end this chapter by exploring the challenges and limitations of behavioural change, based not only from our research on transport, but also within the context of our broader research into the promotion of environmental practices. Most importantly, we want to emphasise the political context within which we are writing. As we noted at the start of the chapter, had we been writing a book on transport and mobility 20 or 30 years ago, we might only have had a footnote explaining the importance of individual behaviour, and perhaps even less about the role of behavioural change. Yet the progressive rolling back of the state (Jessop, 2002) and the changes we've witness in what Giddens (1991) refers to as 'life politics' means that the citizen-consumer (Clarke *et al.*, 2007) has attained a prominence that seems to dwarf all other modes of policy-making when it comes to tackling problems like anthropogenic climate change. This is undoubtedly an ideological process in part; it is about the rendering of individual responsibility to the challenges of our age and about the relegation of the state to a facilitator, a purveyor of 'nudges'. In so doing it has thrust individuals centre stage when it comes to climate change, and yet the epistemological disruption caused by right-wing media outlets (Barr, 2011) has certainly thrown the compelling logic of behavioural change into some confusion when it comes to carbon-reducing practices that directly conflict with the social norms and aspirations of a highly consuming middle class.

Behavioural change is therefore a slippery concept, not only because of its epistemological diversity, but because it is rolled out as a solution, yet it is clearly only one part of the jigsaw of promoting better governance of the environment. We argue here that there are four primary concerns that need to be addressed by both academics and policy-makers if we are to understand what place, and it inevitably does have a place, changing behaviours should hold.

First, as Crompton and Thøgersen (2009) have argued, we need to ask how much change we really think behavioural approaches can deliver. Put simply, it could be argued that tackling anthropogenic climate change requires major structural changes in how we live our lives: investments in new technologies, changes in how urban areas are planned, restrictions on unsustainable practices, alongside changing our own expectations about what we need to live a 'good life'.

Second, it is rare for either researchers or policy-makers to question the logic of behavioural change as an engagement issue. We seem to glibly accept that our role as responsible citizens is to make the changes in our lives that experts and governments believe we should adopt. This is without doubt a highly dis-engaged and dis-empowering form of governance that does not encourage critical questioning of governmental and industry approaches. We simply accept that trying to use the car one day less a week is the 'right' thing to do. But what if we asked about why we use the car, why planners and governments build residential and retail developments that compel us to use cars, or why we cannot insist that public transport is given priority in planning developments over private vehicles? We seem to rarely ask these basic but important questions. As such we have fallen into the trap of working with a definition of sustainable development that is passive and unimaginative, one that is a political fix (Robinson, 2004).

Third, even if we do accept that behavioural change has a role to play, the *modus operandi* used by governments, particularly the UK government, have been driven from a very particular political and epistemological perspective. In this way, the focus has very much been on the individual and the cognitive processes that drive behaviour. Indeed, in adopting this individualist approach, the UK state has adopted social marketing and nudge (Thaler and Sunstein, 2008) as devices to promote particular forms of choice to meet specific policy goals. As such, as Jones *et al.* (2011b) have argued, the UK state has adopted an approach that, on the one hand, wishes to uphold the neo-liberal principle of the free market and choice for consumers, but also wishes to nudge them in particular directions to meet specific goals. In this way 'using the new sciences of choice from psychology, economics and the neurosciences – as well as appealing to an improved understanding of decision-making and behaviour change – a libertarian paternalist mode of governing is being promoted in the UK' (Jones *et al.*, 2011b, p. 15).

Finally, in engaging with behavioural change, we encounter the most basic of questions. This is because we are brought face-to-face with what we think we need to lead a good life. Inevitably, in a consumer society, this usually comes down to having more 'things' to make our lives better, and having more of them this year than we did last year, which we call economic growth. The most fundamental question we need to ask, therefore, is whether consuming and moving as much as we do actually delivers the lives we want. In the context of transport and mobility, what does the hyper-mobile society give us and why do we value it? Perhaps moving less might not be such a bad thing, for as O'Neill (2008) argues 'sustainability can be achieved by taking individuals off the hedonic treadmill to which material consumption is subject and refocusing public policy on those goods that really are correlated with life satisfaction' (2008, p. 128).

If this is the case, perhaps we ought to think about why we need to move so much and what moving less might look and feel like. This is the subject to which we now turn by exploring sustainable place-making.

9 Sustainable mobility

Planning better places to live and (not?) travel

Introduction

Chapter 8 has examined the first of two major approaches towards promoting sustainable mobility that have been adopted among the policy community (a third, which we do not explore in this book, being the development of 'smart' mobilities technology). The focus that has been placed on behavioural change, as Chapter 8 describes, is not merely a function of political faith in citizen-consumers to 'do the right thing', but it is also representative of an underpinning philosophy about small government and the need to exercise neo-liberal principles in all areas of public life. As we argued, this has resulted in a kind of blind faith in behavioural change to act as a panacea for delivering particular policy goals, notably on carbon emissions. Yet we have also argued that this strategy is highly problematic because it is both incremental and expects citizens to act within confined, unambitious and largely unrealistic choice architectures. In other words, put colloquially, why should someone living in a low-density city go through the trial of using expensive public transport that takes longer than a short, private and comfortable trip in a motor car? More to the point, why should that person do so when decades of post-Second World War planning policy have structured places around the convenience of the car?

These are questions we aim to address in this chapter, by focusing on what we argue is the 'elephant in the room': the very spatial configuration of our built environment, which we argue has a profound impact on the way we move in daily life and for tourism and leisure. In doing so, we want to suggest that the issue at hand has less to do with persuading a mass of individuals to change their behaviours through providing them with more and more information about anthropogenic climate change, but rather it is intricately related to how people relate to and connect with the places they live, work and take leisure time. It is, therefore, ultimately about the essence of what it means to dwell in a particular place and what affordances those places give to inhabitants. Are there, for example, safe places to walk, cycle and spend leisure time? Are there affordable, convenient, aesthetically pleasing places to shop? Is it easy to get around through the use of integrated transport? Are the streets, street furniture and architectures ones that encourage dwelling rather than movement?

We want to emphasise in this chapter that it is not our role or intention to pursue a particular type of value-laden mono-prospectus of sustainable urban living for the twenty-first century. Rather, it is about enabling us, all of us, to think critically about what the places around us are like, what affordances they provide for us, how this influences our own mobility and what a change in such living environments might look and feel like in the future. It is therefore a call for engaged thinking about our current living predicament, one that has seemingly cast out a sense that specific places matter. We want to argue that place does matter and by definition we cannot, as white, male, European authors impose a mono-urbanist view for the future. But we can enable more people to question and to critique the assumption that cities and towns have to look and feel as they do.

Accordingly, the chapter begins by examining three symptoms of what we term dystopian mobilities. These are: our obsession with expanding our spatial reach; our consequent neglect of place; and finally our flagrant disregard for a critical politics of place, one that leads us to accept the prescription of the car-based, mobility-intensive society. We then consider three ways in which urban thinkers past and present have explored sustainable mobility. First, we examine the Garden City movement, pioneered by Ebenezer Howard (1898, 1902), in which Howard argued for a reconnection between people and nature. Although his utopian vision of the garden city was evidently based on a suburban model, it was also driven by a concern to reduce the need to travel, by placing amenities and employment close at hand. Second, and in line with a general movement to reinvigorate urban centres, we examine the philosophy and practice of the New Urbanism movement from North America (Glaser and Shapiro, 2001). In efforts to halt the growth of suburban subdivisions, New Urbanism has extolled the benefits of the downtown core, with a move back to mixed-use developments, pedestrian cultures and the relocalisation of food and drink. Third, we explore the recently developed Transition movement, popular in North America and Europe, which has advocated a new approach to sustainable places through focusing on the twin issues of peak oil and anthropogenic climate change as ways of promoting both different forms of planning and changes in social practices. We will examine these models critically, for each has their own benefits and drawbacks. However, we will argue that these places are often more fulfilling and enjoyable places to dwell because of their focus on low mobility.

Finally, the chapter provides two illustrations of sustainable places through focusing on the European example of Freiburg in Germany and the North-American case of Portland, Oregon, in the USA. In doing so, we aim to show how lower levels of mobility and car dependence can be achieved through spatial planning and a consideration of both local culture and topography, as well as some underlying principles of sustainability.

We begin, therefore, with a focus on where we are, drawing on the conclusions from Chapters 2 and 3, in which we argued that we have developed a system of living and moving that is based largely on the motor car, itself currently dependent on a finite resource: oil.

Symptoms of dystopian mobilities: space, place and politics

In his 1994 book, Kunstler (1994) uses an extract from the film *Who Framed Roger Rabbit* to illustrate the extent of lock-in that the car-dependent culture has in North American culture. To take a quotation from the film 'People will drive, Mr Valiant, because they'll have to. And when they drive, they'll have to buy our cars, our tires, our gasoline' (Walt Disney Pictures, 1988).

This quotation provides an excellent illustration of how the ways in which we travel has so much to do with the environment around us. In the case of *Who Framed Roger Rabbit*, this was intimately connected to the construction of the US freeway system and its resultant recasting of US urban space, with attendant shopping malls, strip malls and housing subdivisions, cast out into ever more inhospitable places alongside freeway access points.

We therefore argue that our current mobility predicament comes down to three fundamental issues that are embedded within this example: our (mis)use of space, our relationship to place and our seeming unwillingness to contest these through political action. If we begin with space, it is evident from Chapters 2 and 3 that we have made some catastrophic decisions about how to arrange our living environment. As the example of Exeter, south-west England, made evident in Chapter 2, post-war planning in the UK made a decisive leap towards planning towns and cities around the emergent technology of the car. This had profound impacts on the ways we understand our relationship with space, largely by making places that were previously dis-connected easily accessible, to the point where our expectations of what Banister (2008) terms 'reasonable' journey times exponentially increased and the solution to overcoming the irritation of travelling through geographical space was to be resolved through wider, straighter and faster roads, incrementally added to as congestion increased and speeds reduced. Our first symptom, therefore, is closely related to Banister's (2008) argument that we have come to expect an overall reduction in space through increases in travel time and efficiency.

Second, the failures and consequences of post-war planning have had profound impacts on the places where we live, in part because we have come to expect an ever greater level of access to the world outside. As hyper-mobility takes hold, it is no surprise that the places we neglect are those where we live, because we spend less and less time there. Kunstler (1994) further argues that it is not only the system of roads, shopping malls and subdivisions that degrades place, but it is also the influence of the mono-architectural Modernism that has pervaded so much of the urban landscape, paying little attention to local topography, heritage, culture or human scale. Evidently, this is dependent on national context; the failure of architecture is certainly something felt acutely in North America and arguably in the UK, but perhaps less so in continental Europe. Nonetheless, when a new swimming bath, shopping centre, university research institute or post office is proposed, the architecture seems similar: often out of place and out of scale.

Whatever the diagnosis of place decline, it is without question that for decades many centres in large conurbations have witnessed reductions in inhabitants and have suffered at the hands of a car culture, which has sucked out shoppers to the outer malls and car parks. The resultant urban decline has also been variable, but one thing that does unite both North America and the UK is the emptiness of the city centre, particularly at night, where very few people live and where an often threatening night-time economy has boomed. This second symptom, of place decline, is therefore important because it serves to justify the fight to the suburbs and the malls far away.

Finally, and most potently, we don't seem to care about all this. Eighteen months before writing this book, the lead author spent an evening at a local Transition Town meeting, at which the issue of growth in the city of Exeter was discussed under the banner of 'How big should Exeter be?' The discussion largely focused on the numbers of suburban dwellings that could be spaced here or there and the stress these would place on the city's transport and public infrastructure. Someone representing the local planning authority stood up and presented what they termed a 'vision' for the city, which read like a description of housing areas, ironically with names like 'Greenacres'. One of my colleagues stood up and asked how this in itself could be a vision; surely it was a description of business as usual? And so it was.

We have lost the art and practice of thinking what place should look and feel like. A vision for a place would logically start with what constituted its unique characteristics: its topography, heritage, population, industry, spatial configuration and so on. It would then explore how a vibrant, locally sustaining culture could be derived from the characteristics, one that rested on the assumption that what makes life worth living is our relationships – to people and to place. And that means reconstructing places to fit these characteristics and relationships. We know that this is possible in some way and to anyone who has undertaken social research on travel and transport, it is clear that publics understand this. They bitterly complain about current planning policies that isolate different functions (Barr and Prillwitz, 2014) and argue for better-quality public space and enhanced public transport. Yet nothing changes. This is likely because there is a cultural assumption among most political elites that private is better than public and the individual is more valuable than the collective. As the research by Barr and Prillwitz (2014) highlighted, this has led to a political apathy and fatalism in transport policy that seems hard to break.

The rest of this chapter is about challenging this political apathy. It is concerned with thinking critically about how place relates to mobility. None of the three key ideas nor the two case studies offer a blueprint for any one place, nor do they provide an easy solution, but they do act to challenge us to consider how we may (have to) reconfigure places for a future where the private motor car is not our travel mode of choice.

Better places: back to the future?

The first model of sustainable living that we might look towards necessitates us to cast our eyes back over 100 years. Ebenezer Howard's (1898, 1902) *Garden Cities of Tomorrow* was the primary example of a set of initiatives in the late nineteenth century that characterised interest in improved forms of living arrangement. As Cullingworth and Nadin (2006) and Gilg (2005) note, the nineteenth century was characterised by major concerns over the impoverishment of working-class people in industrial cities, who often lived in squalid conditions with little access to drinking water and sanitation. As Chapter 3 has highlighted, the development of planning healthy places with amenities such as green space and sanitation was a significant precursor to the formalised planning system that eventually emerged in the late 1940s in the UK. Yet long before pioneers like Ebenezer Howard and the Cadbury family in the West Midlands had sought to pioneer major improvements in living standards, either associated with whole new settlements (such as Garden Cities) or the development of towns directly associated with a particular industry (such as the Cadbury family's Bourneville, now in metropolitan Birmingham).

It is without doubt that the lasting legacy of Howard's (1898, 1902) Garden City movement was the most profound of these utopian experiments. In his 1898 book he outlines his argument for creating better places to live and work through the lens of his three 'magnets'. These were the town, the country and then his preferred configuration of the town-country, which combined the benefits of both environments. Howard's argument was that the town, despite its social and economic affordances, was an unhealthy and unfulfilling place to live, with little provision of open space and connections to nature and recreation. Meanwhile, the country was socially isolating and economically impoverishing. By contrast, a combination of the town and country could offer low property rents, an abundance of space, a connection to nature and fulfilling work, while also offering a sense of community, spacious housing and high wages.

Howard had a clear vision not just of what life would be like in these new towns, but also how they would be planned spatially. He envisaged a model that would not be unfamiliar to many geography school students, in which different land uses were allocated according to transport costs and including the necessary productive industries for a city's functioning, such as dairy pasture, fruit and vegetable plots, allotment gardens and small-holdings. He also noted the importance of locally sourcing potable drinking water and of treating the new settlement's sewage. Indeed, he also made provision for smaller, inter-linked settlements. All of these nodes were to be connected primarily by railways (bearing in mind the date of the book's publication) and canals, with little allocation of land area to roads.

In essence, Howard's vision for Garden City living was about what we now call localism; a sense that the model of a town with specific architectural and environmental characteristics, would be an appealing and rewarding place to live and work. It removed people from the squalor of the city and placed them in a 'natural' setting. In this way, Howard's aspirations share a great deal with the North American model of suburbia that Baldassare (1992) has outlined in regard to early metropolitan suburbs like Riverside in Chicago, with its close connection to 'naturalness' (although Howard's vision was very much about working people, not exclusivity).

What makes Garden Cities relevant to the discussion in this chapter is that in very recent times, the idea has gained new impetus through a recognition that there is an urgent need to build more houses and to do so not only on brownfield land, but on Greenfield areas well away from existing urban boundaries (BBC, 2012; *The Telegraph*, 2014). Indeed, in its recently launched prospectus for locally led Garden Cities, the UK government has argued that it aims to offer:

> … a broad support package that Government will offer localities which are ambitious in terms of scale and delivery, and set high standards for design, quality and the provision of green space. We want to encourage people to think in a new way about how they can meet their housing needs. Building on the historical Garden Cities concept, and a legacy of new town development we can be rightly proud of, we want to support localities in delivering inspirational new Garden Cities fit for the 21st century.
>
> (DCLG, 2014, p. 3)

This renewal of the Garden City movement in the UK is a direct response to the need for more housing, but it is also about a recognition by government that constantly adding housing estates to the edges of cities and towns puts stress on existing infrastructure and labour markets, alongside the promotion of a mono-architectural and characterless environment that promotes mobility (Lock, 2012). Accordingly, contemporary Garden Cities are also about a broader agenda that seeks to reduce the need to travel (Banister, 2008) and as a consequence to reduce transport-related carbon emissions.

In a special edition of the Town and Country Planning Association's magazine in 2012, Henderson and Lock (2012) set out their view of what should characterise a Garden City of the twenty-first century, focusing on:

- A strong vision and community engagement;
- Community land ownership;
- Mixed home types;
- Local labour opportunities;
- A distinctive and high-quality architectural design;
- Generous amounts of green space;
- A locally vibrant retail offering;
- Integrated and sustainable transport systems.

In this way, Garden Cities are envisioned to be a way of capturing the benefits of urban living with the affordances of open space and a focus on local sustainability through providing services at a walkable scale. Yet Garden Cities are highly controversial (*The Telegraph*, 2014). First, they require large amounts of greenfield land, well away from existing settlements, to become autonomous and identifiable communities, and as such they have attracted hostility from interest groups such as the UK's National Trust, who in responding to initial discussions about Garden City renewal in 2012 argued that:

> The National Trust believes that land is a precious resource and must be used and managed sustainably to produce the greatest public benefit. As a nation we must be careful to safeguard the productive capability of land for future generations across a range of areas: water, carbon, soils, biodiversity, development, recreation, culture and heritage, and food. Any development on this kind of scale will need to respect this fact if it is to deliver the kinds of benefits Mr Clegg [UK Deputy Prime Minister] talks about, without destroying the countryside.
>
> (National Trust, 2012)

Second, Garden Cities do not necessarily provide a solution to the challenges of urban decay that has blighted much of the developed world since the emergence of auto-mobility culture (Glaser and Shapiro, 2001). Indeed, they are based on a model of economic development that seems to be disconnected to the major shifts towards globalisation that we have witnessed in recent years (Dicken, 2007), which have constructed an economy around the movement and outsourcing of materials and labour across space that make the challenge of relocalising economies highly challenging. To this extent, we must question the level to which the Garden City model can provide a workable approach to reducing movement for work, if not leisure time. Nonetheless, its principles are ones that we shall revisit when we explore the most recent manifestation of sustainable places in the form of Transition Towns.

The emergence of New Urbanism

Before examining the recently formulated Transition movement, we must, however, consider the second major and long-lasting approach to building sustainable places, the so-called New Urbanism (Glaser and Shapiro, 2001; Helbrecht and Dirksmeier, 2012). As early proponents of this avowedly North American concept have argued (Katz, 1994; Langdon, 1994), New Urbanism is based on continental European urban principles, which have successfully integrated living space within the setting of an urban core. In other words, New Urbanism seeks to partly emulate the vitality and mixed economy of European city centres, which combine multiple functions in single streets – residential, retail, hospitality, mobile and leisure spaces all organised into places that seem to work. Accordingly, New Urbanism in a North American context stresses the notion of urban renewal. Unlike the prescription of the Garden City, this concept is interested in regenerating and reinvigorating urban space. It is therefore partly a historical vision of what Kunstler (1994, 1998) refers to as Main Street America, as well as an acceptance that with strident moves towards suburbanisation, there also needs to be something new to offer those citizens who might wish to re-enter the city limits to live and work (Sadik-Khan and Solomonov, 2016).

In this particular North American model of urban renewal, there is also a broader social agenda at work. Sociologists have long studied the impacts of sub-urban living on the communities that come to live in these spaces between the city and the country and Baldassare (1986, 1992) has provided a compelling cri-tique of suburban culture (see Chapter 2) that has challenged the dominant view of the US suburb as one that characterises the American Dream (Baritz, 1989). The essence of New Urbanism is therefore a call to re-scale communities around what many have argued is a human scale (Dagger, 2003; Kunstler, 1994, 1998; Lupi and Musterd, 2006) that could prevent many of the negative consequences of suburban living, such as social isolation, the marginalisation of young people, time-based and financial stress on family relationships, and the development of a hyper-individualistic culture formulated around a cycle of long commutes, work and consumption-based leisure pursuits. As Gorringe (2002) has noted, there is a clear link between urban design and how humans are able to function and relate to each other through the provision of public and open spaces and planning at a scale that affords human interaction and belonging.

New Urbanism has therefore captured the imagination of many in the North American planning community, seeking to advocate a progressive form of place-making. As the case studies at the end of this chapter will examine, there are different forms of New Urbanist architecture and scale, but perhaps the most well-known example is to be found in Portland, Oregon, where there have been strident attempts to develop walkable, high-quality spaces into the historic down-town (Figure 9.1), a model that very much follows the tradition of European city dwelling and which will be explored in the other case study outlined in this chap-ter of Freiburg, Germany.

Figure 9.1. New urban planning in Portland, Oregon.

Source: Reproduced with the kind permission of Michael Mehaffy.

As Sadik-Khan and Solomonov (2016) have documented in the case of New York, the materiality of urban spaces and how this links to human behaviour has become a critical component of New Urbanism, often representing political struggles for space. These struggles become focused on particular elements of the landscape, such as bike lanes, space for seating, pedestrian crossings and the like. Importantly, Sadik-Khan and Solomonov (2016) highlight the ways in which twentieth-century US space has been defined by the motor car through a range of practices. These can relate to how official guides to street design factor-out people in favour of traffic flow, and how pedestrians and cyclists are often regarded as hazards to be removed from streets through the creation of illegal behaviours like jaywalking. In order to reclaim streets for people, Sadik-Khan and Solomonov (2016) explain an impressive range of examples in New York City where changes in street architecture, road layout, crossing points and bike lanes have all increased walkability, cyclability and the propensity to dwell. Such approaches demonstrate, in their view, how relatively inexpensive adjustments can fundamentally shift the balance from a car-dominated streetscape, to one populated by people. As such, New Urbanist principles are applied to achieve a range of goals, from reducing traffic congestion and reliance on motor vehicles generally, to improving the quality of life for residents and supporting local businesses.

New Urbanism, despite its focus on a supposed successful model from European urban planning, is not without its critics. We can characterise the critiques levelled in three ways. First, as Glaser and Shapiro (2001) have argued, despite the initial burst of enthusiasm that met New Urbanism's arrival, it has not significantly dented the aspirations of most Americans to continue to search out cheap, affordable suburban homes well away from the urban centres that New Urbanism so covets:

> We therefore think that the fundamental lesson of urban growth in the 1990s is the remarkable continuity of urban growth patterns. In the last decade, as in all previous post-war decades, urban growth was driven by the increasing importance of consumers and their tastes for cars, good weather, and the skill base of the local community.
>
> (Glaser and Shapiro, 2001, p. 23)

Accordingly, as the continued decay of cities like Detroit highlights (BBC, 2013), there is only evidence that for most US cities, the spiral of decline in urban centres is continuing as cities like Las Vegas continue to grow through the subdivision of outer lands for housing development (Las Vegas Review Journal, 2014). However, this is not a universal process and recent reports indicate that in more progressive planning states like California (*Los Angeles Times*, 2014), increases in population are being accommodated by net increases in apartment blocks and multi-family dwellings, which in some cases have resulted in the demolition of large single-family dwellings. Nonetheless, there is scant evidence to suggest that for the majority of (North-)Americans the culture of a single-family dwelling on a quarter-acre plot and two cars is any less of a dream now than it has been for nearly 100 years.

A second challenge to New Urbanism has been the assumption, very much embedded in the wider literature on planning and urban design, that pedestrian-planned spaces will encourage fewer trips by car and more walking and cycling. In a study of Portland neighbourhoods, Lund (2003) examined the extent to which urban forms and design influenced behaviours on the one hand and the role of personal attitudes on the other. His tentative conclusion was that neighbourhoods that integrated forms of New Urbanism were more likely to be populated by those who held positive attitudes towards non car-based travel and who were looking to live and work in smaller, pedestrian-based communities, indicating an important element of self-selection. Accordingly, his research raised the question of whether such urban forms have a long-term and widespread appeal for the future, or whether a broader shift in social attitudes is required first.

Finally, Talen (1999) has tackled a further orthodoxy of New Urbanist thinking by questioning the extent to which such environments are able to produce or promote a greater 'sense of community'. He characterises the orthodoxy of New Urbanism thus:

> New urbanism, an umbrella term encompassing 'neotraditional development' and 'traditional neighbourhood design' is a planning movement which is gaining increasing popularity. Its promoters stress the conviction that the built environment can create a 'sense of community', grounded in the idea that private communication networks are simply no substitute for real neighbourhoods, and that a reformulated philosophy about how we build communities will overcome our current civic deficits, build social capital and revive a community spirit which is currently lost. Accordingly, new urbanists assert that the main defect of standard suburban development is not aesthetic or even environmental, but is its insidious social effect.
>
> (Talen, 1999, p. 1361)

He questions this orthodoxy through posing a series of questions about both the evidence base for a so-called sense of community and the essence of what such a community might look and feel like. In the first instance, therefore, he argues that much of the evidence on which scholars promoting New Urbanism draw upon to argue for a greater sense of community in renewed neighbourhoods is based on anecdotal evidence and has not been subject to the kinds of rigorous interrogation that would be needed to adopt this approach as a long-term planning tool. However, Talen's (1999) argument is also based on a broader critique of how community is constructed as a simplistic and all-encompassing term by New Urbanist thinkers. For example, he argues that building a sense of community often relies on resident homogeneity and runs the risk of generating the kinds of social isolation that have been evidenced in so-called 'small town' America. Further, he argues that forms of place-based living are, in fact, a traditional view of social association that does not take account of the other ways in which people associate, through interest groups, social media, working relationships and through the integration of travel into daily and leisure patterns. In short, he argues that New Urbanism is problematic because it takes for granted too many assumptions that, at the very least, need further scrutiny:

> The theoretical and empirical support for the notion that sense of community (particularly its affective dimensions) can be created via physical design factors is ambiguous at best. New Urbanism is supported by the fact that research demonstrates a link between resident interaction and environment, and therefore the correlation between public/private space integration and resident interaction is sustained. But to move beyond interaction towards the affective dimensions of sense of community is problematic since the effectuation of a sense of community in these terms is usually only achieved via some intermediate variable (for example, resident homogeneity and affluence). This leaves open the question of whether or not any number of other design creeds could produce the same result via a different design philosophy.
>
> (Talen, 1999, p. 1374)

Transition Towns

The Garden City and New Urbanism models of development provide contrasting visions and solutions to the challenge of the automobile society. One the one hand, the Garden City movement is about the development of greenfield land that fosters new forms of community and, through the greater affordance of open space and access to the countryside, a connection to nature that is deemed to benefit human well-being. The New Urbanist prescription is far more concerned with redeveloping and rejuvenating decaying urban centres and is founded on a partly historic notion of the typical American Main Street within a small town, where high density is regarded as a way of suppressing mobility and creating community. As we have already noted, neither of these prescriptions deal adequately or critically with the notion of what constitutes community and how belonging can also lead to geographies of exclusion (Talen, 1999, 2001).

Into the gap between these two models of low low-mobility development has emerged a third and much more recent idea, that of the Transition Town. Its recent emergence is partly a function of its twin concerns: peak oil (Deffeyes, 2001; Heinberg, 2007) and anthropogenic climate change (Hopkins, 2008). Although much has been written about the Transition movement already (Aiken, 2012; Bailey *et al.*, 2010; Haxeltine and Seyfang, 2009; North, 2010; Seyfang and Haxeltine, 2012), our concern with this movement is about what it has to say concerning the development of sustainable, low-mobility community and the transition away from a carbon-dominated economy.

Transition is founded on the notion of resilience (Adger, 2000), which is a concept drawn from the discipline of ecology. In drawing upon resilience, the acknowledged founder of the Transition movement, Rob Hopkins, has argued that the dual challenges of peak oil and anthropogenic climate change (discussed in Chapter 7) present a major threat to communities that are founded on a model of consumption that necessitates the transportation of vast quantities of material, food and energy to service their needs. In invoking the resilience concept, he argues that during periods of oil price rise (particularly relevant when he authored the *Transition Handbook*), supply shortage and the potential disruption caused by weather events associated with climate change, communities are inherently vulnerable because of their reliance on globalised trade networks and their inability to live according to a reduced-consumption model (Hopkins, 2008).

Accordingly, Hopkins argues for a focus on resilience, which is founded on the notion of 'bounce back' (Adger, 2000; MacKinnon and Derickson, 2013). In other words, what are the ways in which communities can insulate themselves against the shocks that may occur from a changing economy, being ever more framed by shortages in fossil fuel resources? Hopkins' (2008) approach is partly a pragmatic reflection of what Heinberg (2007) and Kunstler (2005) have explored in theory, which is a gradual but steady rise in energy prices that affects not only domestic fuel prices but the ability of communities as they are currently configured to maintain economic stability – be that through long commutes or the filling of supermarket shelves with everything from Kenyan beans to Brazilian bananas.

Hopkins (2008) provides some useful examples of what community resilience in this context might involve (Table 9.1). In essence, resilience can be interpreted as two dependencies. The first of these is mobility. As noted previously, place-based communities are increasingly reliant on oil-based transport to supply

Table 9.1. Forms of community resilience

Not Adding Resilience	Adding Resilience
Centralised recycling	Local composting
Ornamental tree plantings (e.g. Millenium Forests)	Productive tree plantings
Sourcing organic food internationally	Local procurement specifying local production, supporting emerging and new industries
Imported 'green building' materials	Specifying local building materials (cob, hemp, etc.)
Low-energy buildings	The local Passivhaus
Carbon offsetting	Local community investment mechanisms
Ethical investment	Local currencies
Buying choral CDs	Singing in the local choir
Sky sports	Playing football
Consumerism	Reciprocity

Source: *The Transition Handbook* by Rob Hopkins, published by Green Books, www.greenbooks.co.uk.

foodstuffs using distribution systems that are highly efficient, which means they are very cost-effective. Yet this also introduces a high degree of vulnerability into such systems, which are reliant on regular, low cost and un-interrupted shipments of goods to supermarkets. Quite apart from the exotic locations from which much of our food originates, our daily foodstuffs like milk, bread and perishable produce all largely travel long distances by road transport.

A second dependency is much more profound. This is the psychological dependency we have on these products and their immediate availability, 24 hours a day. We are nearly all entirely dependent on a fragile and very rigid distribution system to provide the basics of our everyday living – milk, bread, vegetables, fruit, meat and fish. Any breakdown in the system can and would cause social unrest and psychological stress, as evidenced when there have been breaks in drinking water supplies due to natural disasters.

Hopkins (2008) argues that the impacts of such stresses can be reduced by promoting greater community resilience through both relocalisation and greater community capacity-building. As Table 9.1 highlights, this might be envisaged by a community agriculture project, which serves the dual purposes of sourcing local food with no need for long transportation and building a sense of pride and belonging among participants. Indeed, as Hopkins notes, resilience requires us to reconnect with places through reappropriating activities that we have often outsourced. A good example of this might be recycling household waste; in most cases, householders simply collect and 'put out' recyclable materials for collection by the local authority. However, a resilient system might attempt to involve people in the collection, sorting and redistribution of such materials, thus building social capital and contributing to local resource gain. In this way, Transition, through resilience, seeks to connect systems and people through changing accepted social practices.

Transition, as Barr and Devine-Wright (2012) have argued, is therefore about bridging the gap between individual change on the one hand and restructuring towns, villages and even cities, on the other. The essence of Transition is about what Bailey *et al.* (2010) have referred to as relocalisation. In this way, Transition is about sourcing and managing resources for daily life within specific geographically defined areas, which leads to a series of approaches that involve localised food growing and procurement, energy production that utilises low carbon and locally specific sources (wind, solar, hydro-electric and marine) as well as strategies for diverting materials from landfill sites.

Apart from what are often very pragmatic approaches to reducing carbon emissions and increasing resilience, Transition also focuses on underpinning issues of consumption (Barr and Devine-Wright, 2012) and has done so through both a focus on what Hopkins (2008) refers to as inner-transition (the changing of the self and one's approach to life satisfaction through consumption). In so doing, it is argued that individuals need to reflect and change their own practices as a way of recognising the value of community as a key unifying mechanism for increasing well-being (Aiken, 2012). This has led to an upholding of 'community' as a central pillar of the Transition movement. As Aiken (2012) has highlighted, this has not been theorised to any great degree, but is founded once again on the notion that better places to live are based on a set of characteristics that uphold place specificity. Within Transition, this also comes with a set of values that stress the importance of consensus decision-making, rather than relying on conflictual models of political behaviour.

As Bailey *et al.* (2010) noted, Transition is a new movement that doesn't necessarily have an overarching vision of what places could or should look like in the future, but it does argue that the emergence of a relocalised economy is one that is founded on the principles of reduced consumption, consensus decision-making and the aspiration of resilience. It is too early to state what the likely success of such a movement will be or whether indeed it will have the longevity of North America's New Urbanism and Europe's Garden City movement. What is likely, is that attempts to plan places that are sustainable in a holistic sense will be difficult, require considerable investment and a long-time commitment to see them through. It should not be forgotten that in most parts of the developed world, the built environment has been constructed to service the car and up-rooting such a literally embedded form of mobility will be hard and may be something forced upon us by the changing resource-base on which driving depends.

Yet there are examples of where places have been planned that seem to meet the requirements of sustainability and have successfully suppressed the need and desire to travel using private motor transport. In the final sections of this chapter, we explore two well-known case studies of sustainable places, from Europe and North America.

Promoting sustainable mobility

If we want to move beyond the specific doctrines of the Garden City Movement, New Urbanism and Transition as particular and often prescriptive forms of place-making, we should first consider some essential principles that each of these particular interpretations on 'making places' brings to a wider discussion of planning sustainable places. Of all the scholarship in this field, perhaps the most lucid approach has been provided by David Banister (2008) who has plotted his sustainable mobility paradigm (Table 9.2) in contrast to conventional approaches towards transport planning. As we noted in Chapter 7, the evolution of sustainable transport policy in countries like the UK has certainly attempted to deal with some of the environmentally negative consequences of transportation. However, this has not led to a step change in the ways in which we have attempted to deal with other sustainability aspects, such as economic, social and cultural elements. As such, the shift towards more sustainable transport policies has largely been instrumental in nature; that is, the changes have responded to particular pieces of EU and UK environmental legislation to reduce noise, air pollution and carbon emissions. Yet to promote sustainable mobility we must consider how we can make changes in the underlying socio-economic structures that govern our mobility, which partly comes down to how we plan our towns and cities, and also relates to how we use these through our everyday practices.

Table 9.2. Banister's (2008) conventional and sustainable approaches to mobility

The conventional approach – transport planning and engineering	An alternative approach – sustainable mobility
Physical dimensions	Social dimensions
Mobility	Accessibility
Traffic focus particularly on the car	People focus either in (or on) a vehicle or on foot
Large in scale	Local in scale
Street as a road	Street as a space
Motorised transport	All modes of transport often in a hierarchy with pedestrian and cyclist at the top and car users at the bottom
Forecasting traffic	Visioning on cities
Modelling approaches	Scenario development and modelling
Economic evaluation	Multicriteria analysis to take account of environmental and social concerns
Travel as a derived demand	Travel as a valued activity as well as a derived demand
Demand based	Management based
Speeding up traffic	Slowing movement down
Travel time minimisation	Reasonable travel times and travel time reliability
Segregation of people and traffic	Integration of people and traffic

Source: Banister (2008) adapted from Marshall (2001), Table 9.2.

Banister (2008) therefore argues that a sustainable mobility paradigm isn't simply about thinking about transport planning, it is about planning places as sites of practices, which respond to people's needs and seek to shape their mobility practices. Accordingly, sustainable mobility for Banister is a holistic approach that links transport to other facets of daily life, both in structural terms and at the micro level of analysis. Table 9.2 presents his sustainable mobility paradigm alongside conventional approaches to transport planning. What is noteworthy from this table is how a number of contrasts emerge when reading across the table; from a focus on transport, we move to a focus on the social; from a focus on structures, we shift towards exploring people; from understanding the street as a road, we shift to interpret the street as a public space.

These guiding principles are close to but not the same as the prescriptive and often restrictive or place-specific interpretations that New Urbanism or the Garden City movement have brought. Rather, they are about getting planners, businesses, home builders, local governments and people to think differently about what purposes places serve and what role movement has within them. Most importantly, Banister's (2008) agenda is about up-turning the focus in transportation geography and planning from asking the question 'how can people move at greater speed and with more efficiency?' to 'what kinds of places do we want to live and work in?' While mobility inevitably has to be part of the answer to this second question, it is not at the heart of it. Thus, the focus shifts towards making and sustaining places that work for people, where mobility is integrated into social practices and physical fabrics that serve people's well-being.

Sustainable mobility therefore introduces a set of principles, not rules, about how we might do things differently. In a specific place, at a specific time, there will be different interpretations of sustainable place-making. This might result in a range of outcomes; for one place it might be pedestrianisation, retail rejuvenation and local employment investment; for another it might be a light rail project, cycle scheme and providing access for more people to public transport. Most importantly, sustainable mobility is about ensuring that places can continue to be vibrant, economically viable and accessible. We now turn to two examples of how sustainable mobility has been incorporated in both US and European contexts.

Sustainable place-making – the case of Freiburg (Germany)

As the title of Buehler and Pucher's (2011) paper implies, the German city of Freiburg has attained the status of Europe's 'environmental capital'. Situated in the Black Forest, near the south-western German border with France and Switzerland, Freiburg is a moderate-sized city of approximately 230,000 inhabitants and is well-known for being a major university city. However, it is perhaps best known for the radical changes that have occurred since the 1970s in the way that transport has been managed and the promotion of pedestrian, cycling and the use of mass transit has been developed. FitzRoy and Smith (1998) state that between 1983 and 1995, the number of trips on public transport in Freiburg rose from 27.7 to 95.9 million, an annual increase of over 75 per cent. This was despite only moderate population growth overall in the city. As Ryan and Throgmorton (2003) note, this significant change can be traced back to changes made initially in the management of transport beginning in the 1970s with significant investment in the city centre to try to reduce the number of families and retailers relocating to the urban fringe. Noting an important environmental discourse that emerged in the city as a result of the threats posed by acid rain, other initiatives soon followed, such as the pedestrianisation of the city centre and investments in mass transit and the remodelling of streets to de-emphasise the importance of motor vehicles over bicycles and pedestrians. However, Medearis and Daseking (2012) also identify some other, broad underpinning themes that have shaped Freiburg's reputation as a German and European leader in positive planning and low-carbon mobility. First, Freiburg was carefully rebuilt after the Second World War, when almost 85 per cent of the city centre was destroyed in the conflict. Unlike the often brutalist and modernist approaches adopted in the UK, planners carefully and attentively reconstructed the city centre according to the pre-war street plan and also ensured that the architectural integrity was maintained between pre- and post-war building. Second, in line with appropriately maintaining the scale of development, Medearis and Daseking (2012) note the importance in regional German planning to the role of landscape and environmental protection, which has had a particular impact on Freiburg by managing the urban forests and limiting development in forest lands immediately outside of the city. Third, there have been major changes in the layout of roads and streets, so that priority is given to cycling and walking, and of course dwelling, rather than motor vehicle movements. Finally, Medearis and Daseking (2012) highlight the long-term nature of change that has resulted in a profitable and sustainable public transit system, which in terms of light rail is accessible within 500 metres to 70 per cent of the population. Only with long-term, sustained and continued commitments to investment, they argue, has Freiburg managed to significantly reduce car use and promote other modes of travel.

Kenworthy (2006) uses Freiburg as one example among several of how transport planners can radically change the mobility practices of urban dwellers. He highlights ten key factors that can lead to reduced reliance on private motor transport (pp. 68-69):

- The city has a compact, mixed-use urban form that uses land efficiently and protects the natural environment, biodiversity and food-producing areas;
- The natural environment permeates the city's spaces and embraces the city, while the city and its hinterland provide a major proportion of its food needs;
- Freeway and road infrastructure are de-emphasized in favour of transit, walking and cycling infrastructure, with a special emphasis on rail. Car and motorcycle use are minimised;
- There is extensive use of environmental technologies for water, energy and waste management – the city's life-support systems become closed-loop systems;
- The central city and sub-centres within the city are human centres that emphasise access and circulation by modes of transport other than the automobile, and absorb a high proportion of employment and residential growth;
- The city has a high-quality public realm throughout that expresses a public culture, community, and equity and good governance. The public realm includes the entire transit system and all the environments associated with it;
- The physical structure and urban design of the city, especially its public environments, are highly legible, permeable, robust, varied, rich, visually appropriate and personalised for human needs;
- The economic performance of the city and employment creation are maximised through innovation, creativity and the uniqueness of the local environment, culture and history, as well as the high environmental and social quality of the city's public environments;
- Planning for the future of the city is a visionary 'debate and decide' process, not a 'predict and provide', computer-driven process;
- All decision-making is sustainability-based, integrating social, economic, environmental and cultural considerations as well as compact, transit-oriented urban form principles. Such decision-making processes are democratic, inclusive, empowering and engendering of hope.

Accordingly, cities that reduce their reliance on private motor transport are those which have made a whole raft of changes not only to their infrastructure and technologies, but also to the culture of planning sustainable places, a factor which has certainly been borne out in Freiburg, but also in the following example from the USA.

Sustainable place-making – the case of Portland, Oregon (USA)

Among the growing examples of attempts to create more sustainable places in North America, one illustration stands out from the rest. The City of Portland in Oregon has for many years been regarded as a trail-blazer for developing innovative planning approaches to reduce the impacts of urban sprawl (Abbott, 1997). As the City of Portland (2016) has noted in its most recent draft plans for the city, contemporary urban areas face a whole raft of challenges, not simply confined to urban sprawl or congestion, but also the impacts of economic downturns, social mobility, well-being and physical health of residents. In this sense, a look at the City of Portland's planning website (www.portlandoregon.gov/bps/50531) highlights the ambition Portland's city government has for both the central city and its surrounding counties.

Portland's approach has long been associated with adopting the principles of New Urbanism (Abbott, 1991, 1994), in particular the adjustment of zoning laws to promote higher density, mixed-use environments, alongside vigorous investments

Figure 9.2. New urban planning in Portland, Oregon.

Source: Reproduced with the kind permission of Michael Mehaffy.

in public transit and the development of pedestrian and cycling corridors (Abbott, 1991). Figure 9.2 is a useful illustration of the ways in which formerly auto-dominated streets have been remodelled to make them inviting and accessible to city dwellers and mixed-use activities. In this way, Portland's city government have attempted to both rejuvenate existing urban areas, while promoting higher density environments linked to rapid transit, to avoid the emergence of single-occupancy-based sprawl.

In exploring some of the principles within Portland's approach, the recently published Central City 2035 Plan (part of the overall Comprehensive Plan) provides some useful examples of the kinds of foresight being brought into play over the coming decades (City of Portland, 2016). At the centre of this plan is a bold but critical statement and working assumption: 'Healthy cities need healthy hearts' (City of Portland, 2016, p. 9). As noted in the document, it is the intention of the city government that much of the growth that will occur in the city over the coming decades will occur in the central core, rather than the outskirts. As such, the Central City 2035 Plan seeks to set a framework for this. Specifically, the city government has outlined six key areas for focus as the planning process moves forward:

- Celebrating Portland's civic and cultural life through further improving sites of cultural interest and interaction;
- Fostering creativity and innovation through linking creative industries with universities;
- Developing the Willamette River frontage for civic use, building on its heritage as a vital communications network for the city;
- Developing and designing streets to be high-quality public spaces for interaction and dwelling;
- Developing a corridor of green space, with tree canopies and a high-quality space for commuters to use for walking and cycling to work;
- Increasing the resilience of the central city, through encouraging a greater mix of employment opportunities in the central area.

Accordingly, the Central City 2035 Plan places a focus on ensuring that Portland maintains and develops its character as a vibrant and sustainable settlement. In large part, the academic literature that has been published on the success of this long-term strategy is praiseworthy of such attempts. As early as the 1990s, Abbott (1997) has pointed to the success of Portland's attempts to reduce urban sprawl by reinvigorating downtown areas of the city. Nonetheless, there are those who point to the challenge of actually stopping urban sprawl and reliance on single-person vehicles. Both Podobnik (2011) and Song and Knaap (2004) highlight that while newer communities built on the outskirts of Portland are positive in terms of their deployment of key New Urbanist design principles, there is still a major reliance on cars for transport. Indeed, Lund's (2003) empirical study of Portland residents highlighted that in many ways, the sustainability of new settlements was largely dependent on the personal attitudes of residents, rather than being influenced by urban design and form. As such, the evidence is likely to show that despite the major advances made in a city like Portland, there is still a major appeal for the private motor car to be the mode of choice for many residents in this North American city. Nonetheless, this should not detract from what is an impressive set of achievements and the fact that many parts of Portland are highly successful places. And Abbott (1997) has made the point that the success for Portland has not necessarily been in the physical designs themselves, but rather has come about because Portland's planners have widely engaged with residents about that they propose to do 'The organization and character of the public realm are thus the key variables that make Portland different' (Abbott, 1997, p. 45).

This is a particularly important finding because it highlights the need to ensure that the development of places occurs in a democratic and inclusive manner. As we highlighted in Chapter 2, much of the 'freeway thinking' of the post-war period in the UK and North America was dominated by an expert-led, top-down and disengaged manner. Robert Moses perhaps uniquely embodied this approach. Yet in Portland and in many other good examples, city planners are now being led by the people who will live in the places that are being redeveloped. This can have immense power through legitimating approaches and situating them within a context that is meaningful for people's everyday lives.

Conclusion: reclaiming the streets

In Sadik-Khan and Solomonov's (2016) *Street Fight*, they made a compelling and overtly politicised case for reclaiming streets for people, using many examples from New York to illustrate the ways in which urban life has been so skewed towards the demands of traffic flow and speed. This chapter perhaps more than any other begs us to ask the question: what are places for? For many decades we have assumed that places serve particular and singular purposes: they are shopping centres, industrial estates, office blocks, residential suburbs and recreation centres. We have perhaps come to see the world through this lens for some very understandable reasons. First, as we noted in Chapter 2, the zoning of land use was a response to the dirt, pollution and squalor of industrial city cores. Enabling people to move out and take advantage of cleaner air and open space was evidently a good thing. Second, the rapid adoption of the motor car as the preferred mode of transport in the first half of the twentieth century has certainly introduced a progressive automisation in everyday life that privileges private space and seclusion above public space and interaction. Accordingly, in trying to think through what places are for, we need to consider both where we have come from and who we are now. What we find is that to change places, we must change our outlook on what we want from life and how we connect to others around us.

This chapter has presented several ways we might capture that process. However, the thread that ties all of these together (be they Garden Cities, Transition Towns or New Urbanism) is a passion for specific places and an understanding that people are at the centre. Interestingly, we do seem to know how to do this as a society; we can talk to each other without hesitation about what makes a good place: green spaces, walkability, high-quality public transport, independent shops, low commuting distances and spaces for frequent interaction. Yet it also seems to be a fight to deliver these attributes. Perhaps that is because despite what we know to be true, we still end up conforming to a social norm delivered to us courtesy of the major house builders, car manufacturers and supermarkets.

10 Conclusion

What future for mobility?

Introduction: mobility in social context

There are many ways in which a book entitled *Geographies of Transport and Mobility* could examine the processes of movement through space. Our approach has been to emphasise the dynamic nature of transport and mobility as something that is characterised by key geographical concepts. First, as Rodrigue *et al.* (2013) have highlighted, the geography of transport has conventionally been about understanding how people, goods and services are configured and how they move through space. Indeed, transport has often been regarded as a mechanism for overcoming the limitations of distance and we have emphasised throughout this book the ways in which the desire for increasing the speed of transport has been motivated by a concern to reduce the role of distance as a barrier to communication and human interaction. Second, we have highlighted the important role of place in providing counter-narratives to the dominant assumption that speed, mobility and travel in general is a good thing and that hyper-mobility is the ultimate goal of an advanced and enriched society (Banister, 2008). Rather, we have sought to question the underpinning logic of banishing the limitations of distance through speed, instead arguing for an approach that celebrates the importance of place to dwelling, human well-being and improving social interaction. This argument is not intended to eliminate mobility, but it is designed to respond to the call by Banister (2008) for a rebalancing of the relationships between dwelling and mobility, and a recognition that the environmental challenges we are facing and living up to will require a less carbon-intensive society, one that will need to rely less on certain kinds of mechanised transport and learn to live with products and services that have travelled shorter distances.

In emphasising the tensions between the aspiration to conquer space and a celebration of place, we have also used two other key geographical concepts to frame this book's contribution. Third therefore, we have recognised and highlighted the temporal shifts in social practices that have resulted in the contemporary tension between the dominant tendency for hyper-mobility situated in contrast to lifestyles based on dwelling (Sheller and Urry, 2006; Urry, 2007). Since the mid nineteenth century, the affordances of mechanised transport systems (steam trains, electric trolley buses, trams and finally motor vehicles) have enabled a fundamental shift in the ways that we relate to place and utilise space. This was witnessed initially through a reconfiguration of domestic and work spaces, in which living further away from the place of work became a necessity and then a desirable component of the 'good life' (Baldassare, 1986). In the developed world, the enrichment of a growing middle class, and even the working classes, led to greater amounts of leisure time and the ability to travel for paid holidays. At first these were mass movements of the population from factory town to beach resort, but the advent of cheap air travel has subsequently opened up tourism to a vast array of markets (Sheller and Urry, 2004). In these ways, transport and mobility have both influenced and continue to be shaped by the social practices that underpin the way we choose to consume.

Finally, we have implicitly used the geographical concept of scale throughout this book. Scale manifests itself through each of the other three concepts we have described. It is inscribed through our temporal relationship with transport and mobility and enables us to appreciate how we have shifted from place-based communities of towns and villages in the early nineteenth century to globalised communities of practice and interest in the twenty-first century, connected through both the technologies of physical mobility and those of virtual communication (Sheller and Urry, 2006; Urry, 2007). Scale is also implicated in the dilemma we have described between our desire to 'consume' more and more places through shortening distances and compressing space and the recognition that slowing movement down and creating better places to live and work may be both desirable and necessary in the coming decades (Banister; 2008; Dickinson and Dickinson, 2006). We argue here that as societies and individuals, we have fundamentally become 'out of scale' in the ways we choose to live our lives and that this is likely to become a major challenge in a century that will be defined, however it is manifested, by climate change and the challenges associated with reducing carbon emissions. Ultimately, this may mean a rescaling of transport and mobility that makes dwelling the norm and mobility the exception.

Our task in this book has been to explore how these geographical concepts are so critical to an understanding of transport and mobility in the twenty-first century and to highlight how it is only possible to study these issues with reference to broader social and economic processes. Accordingly, we need to view transport and mobility in context – through the shaping of changing economic processes of production and consumption, in the light of social and political change and liberalisation, and in the face of rapidly developing technologies that open up the possibilities for both physical and virtual travel (Cresswell, 2010, 2011). This means situating transport and mobility as fundamentally social constructs and treating our relationship to them as socially embedded rather than part of separate and conscious 'transport decisions' (Freudendal-Pedersen, 2009). In this final chapter, we explore the academic and practical implications of adopting such an approach and the likely future of what has been a momentous but short period in human history, where we have been able to travel freely and at relatively low cost. We will ask what society might look like if this period of hyper-mobility came to an end and how our human consciousness could cope with becoming people of dwelling rather than people of travel.

A geography of dwelling?

Implicitly, this book relates to the geographical context of movement, of not being sedentary, but regularly engaging in travel, tourism and even relocation. Yet as transport geographers we rarely discuss geographies of dwelling and their connection to how we plan places and what kinds of practice promote or suppress the need to travel (Blunt, 2005). We want to argue here that researchers need to place much greater emphasis on dwelling as a key construct for making the aspirations of sustainable mobility a more realistic vision. This is not to advocate a specific framing of dwelling as it has been advocated by different promoters (e.g. the New Urbanism of North America) (Katz, 1994), but rather to highlight the need to critically engage in academic research on *lower* mobility, and *slow* travel lifestyles and places (Dickinson and Dickinson, 2006).

In making this call, we argue for a reorientation by researchers that explores the properties and prospects of low-mobility lifestyles and as such we call for scholarship that critically engages with the likely forms of built environment, technology, economic processes and potential behavioural changes that would be required to move towards the political goal of Banister's (2008) sustainable mobility paradigm. This task has already begun in some areas of transport geography and in the sociology of mobilities. Dickinson and Dickinson (2006) and Lumsdon and McGrath (2011) have highlighted the role of slow travel and its relationship to broader lifestyle choices and social relations. Indeed, in the field of behavioural change, there has been considerable scholarship that has examined the potential for attempting to encourage individuals to reduce their travel footprint through tele-working and car-sharing (Harker Martin and MacDonnell, 2012; Rietveld, 2011; Steininger and Bachner, 2014). However, these studies have largely been focused on the ways in which people construct and deploy transport-focused decisions within daily travel or tourism settings and they tend to be focused more on the individual as a unit of analysis. What we wish to argue here is that the scale of study also needs to explore the underlying social trends that promote mobility. In this regard, we argue that transport and mobilities researchers need to look towards two powerful intellectual agendas. First, there is a need for scholars to engage more readily with the ideas of social practice theorists, which have begun to make an impact within tourism research (Verbeek and Mommaas, 2008). As we have outlined in Chapter 8, social practices are understood as broad, habitual and temporally shifting sets of shared behaviours that are heavily dependent on physical and technological settings. What makes their study important is that they enable what appear to be individual behavioural decisions to be placed into a broader social context and they help us to appreciate why it is so difficult to change behaviours.

Although transport and mobilities researchers have engaged with this field somewhat in the past (Verbeek and Mommaas, 2008), it is without question that the vast majority of research using social practice theory has been related to domestic practices such as energy use or water consumption (see Shove *et al.*, 2012). Yet mobility is perhaps the quintessential social practice, because it is something that is the result of the interaction between the physical environment, technology and social norms. The kinds of research questions that transport and mobilities researchers need to address in this regard relate to how mobility practices are initially formed and how these practices change through time and space. What, for example, are the processes that lead to the development of particular forms of tourist practice, such as the so-called 'gap year' travel experience of teenagers or the development of short-haul city breaks provided by low-cost airlines (Graham and Shaw, 2008)? Such practices have rapidly developed and it is likely that they are the complex out-turn of economic processes (such as airline deregulation and continued shifts towards a Post-Fordist model of tourism production), developments in technology (such as more efficient aircraft) and changes in social attitudes (such as the development of a mobile group of newly retired individuals). By looking at the example of tourism mobility in this context, it is clear to see that focusing on one aspect (such as behavioural change) becomes challenging given the number of influences involved.

A second key intellectual agenda to consider is the role of urban studies and planning and how we can come to understand the connections between dwelling and mobility. As Cullingworth and Nadin (2006), Gilg (2005) and Knowles *et al.* (2008) have noted, the links between planning, urban design and transport have always been problematic, not least because the construction of roads and freeways has often been in conflict with the fabric and historic design of urban centres. Yet moves to regenerate North American cities (Glaser and Shapiro, 2001) have made explicit the links between mobility and urban decay that have led to a re-evaluation of how planning relates to transport in the USA and Canada. Nonetheless, as we discussed in Chapter 9, New Urbanism, or for that matter the European Garden City movement, does not provide a blueprint or clear pathway for understanding the links between urban form and mobility and what makes a good place to live and dwell in the twenty-first century. This is because these are both prescriptions designed to deal with particular situations and reflect specific values about how places should look. Indeed, they are, to some extent, fundamentally different because of the different emphasis they place on urban regeneration and new town (greenfield) development. Rather, we are interested in understanding how the places that people live in now can be better and more liveable, and the processes through which this might be achieved.

What this means is drawing upon a new literature that upholds the importance of the 'local' and the 'home' as a site for co-producing better places to live (Blunt and Dowling, 2006). In part therefore, transport and mobilities researchers should be drawing on the various threads of research in human geography that have explored the cultural geographies of the home and of dwelling (Blunt, 2005) to explore how a sense of home is co-constructed and develops, and what conditions may present this from occurring. This is critical because:

> For many people, home is a place of belonging, intimacy, security, relationship and selfhood. Through their investments in their home people develop

their sense of self and their identity. Others experience alienation, rejection, hostility, danger and fear 'at home'. Houses are the material structures that provide the scaffolding for emotional investments, social relations and meanings of everyday life.

(Dowling and Mee, 2007, p. 161)

Accordingly, if we accept the assumption that promoting dwelling can reduce at least some of the desire for hyper-mobility, then an understanding of what makes a good place to live is critical and it is incumbent on transport and mobility researchers to join the dots between research on geographies of dwelling and travel decision-making and practices. And this is where there is a second opportunity for radical thinking on how we, as academics and practitioners, relate to the publics who live and work in specific places. We argue that rather than adopting traditional methods of knowledge construction about what enhances belonging, place attachment and dwelling, we need to co-produce understandings of place that begin to challenge expert-led assumptions about urban design. This is partly in response to Forester's (1999) long-established call to invoke participatory and politically active planning discussions in local communities to provide democratic and inclusive methods for rescaling places according to local need. Here we advocate extending the notion of participatory planning through the lens of knowledge co-production that is not just about an extended form of consultation or deliberation, but it is about challenging the 'normal' (Lupton, 2013) ways in which we go about doing science and social science. In exploring a post-normal form of science, authors such as Whatmore and Lane (Lane *et al.*, 2011) have advocated the use of non-standard research methods to explore what they term knowledge controversies through multiple stakeholder 'competency groups', in which different kinds of knowledge and expertise are used to understand a controversy (Barr, 2017). Such controversies might relate to the attribution of climatic change to weather events or the challenging of scientific methods for reducing natural hazards. But they might equally relate to the conflicts surrounding place-making and place-keeping. It is without question that we currently have very few mechanisms at our disposal to actively engage and enable members of communities to participate in a meaningful way to what their local area should look and feel like in the future. A major outstanding question for researchers is, therefore, how we can use participatory mechanisms to build knowledge frameworks that actually have traction to change decision and counter the dominant economic growth narrative that has for so long privileged the car and suburban living as the only viable model of living and movement arrangement. Indeed, as scholars, we have a duty to address what is arguably a major democratic deficit in decision-making about sustainability (Dobson, 2010; Robinson, 2004).

Changing (transport?) policy

Sustainable place-making and place-keeping are clearly important as exercises in creating valuable places to live. Yet there are limitations, as we have argued throughout this book, in relying on individuals or even communities to change their living arrangements. This has not so much to do with a lack of will, but rather a deficient choice architecture, which does not support such changes. Indeed, even if we were able to create widespread sustainable places, the external economic environment would not necessarily support them.

Accordingly, this leads us to pose some broad and deep questions concerning the wider context of transport policy and how this relates to the incentives that act on individuals, groups and places to change how we travel. We can take the example of retail development as an illustration of this problem. In the UK, the planning system has recently been reformed to make a presumption in favour of development (DCLG, 2012), removing its statutory obligation to promote sustainable development as a first principle. Such a move inevitably reflects both the political values of the current Conservative government, but it also highlights a response to the economic downturn witnessed since 2008 in the UK.

Set against this, the opportunities for change are limited. In a neo-liberal economy that promotes growth above other priorities, there are few arguments that are levelled against the construction of large retail parks on greenfield land on the outskirts of settlements, which are constructed with private finance and which provide employment and contribute to the consumer-led society that so many covet. Yet we know that such developments promote travel and they do so using one form – the private motor car. There are several arguments we can raise against such an approach. In the first instance, there is the generation of traffic that results from such developments. On a local scale, this causes congestion and noise, while contributing to carbon emissions at a global scale through increases in journeys. Second, the habitualisation of driving behaviour becomes ever more embedded through the signification of driving's acceptance in the built environment. In other words, it is a clear statement of what society accepts and promotes as being desirable behaviour – using private transport to access retail environments on the outskirts of towns and cities. Third, the land allocated to such developments is far more than could ever be provided in densely built urban areas and so there is an abundance of choice, central to the idea of free markets, which people have access to. Fourth, such developments naturally draw retailing out of city centres and in doing so they contribute to the decline of urban centres. Indeed, this does not only affect the central core of a town or city, but the suburban shopping precincts and walkable spaces that can act to reduce reliance on the car.

The question of transport policy therefore becomes an issue of principle and a statement of values about what a government or local authority believes are important. As Barr and Prillwitz (2014) have demonstrated in their research on travel practices, a sense of fatalism has crept into British society that mistrusts the ways in which governments attempt to change behaviours, in the main because these are not supported by the investment in infrastructure required to respond favourably. Indeed, as participants in this research highlighted, the pursuit of individualisation within society as a goal of personal betterment has meant that the car, as Rajan (2006) has argued, has become an icon of neo-liberal values.

In this way, it is highly irregular for individuals to be asked to make personal sacrifices in the name of some form of collective citizenship, when neither political philosophy nor practical logistics are supportive of such changes. In other words, the very basis for promoting sustainable travel does not exist; that is, the supporting infrastructure and economic conditions do not support a wholesale rebalancing of travel priorities and so policy has come to settle on an approach that places the blame for excessive car travel at the foot of consumers, who need to make 'better choices'. We argue here that until the political philosophy that does not reward different choices is changed, it is unlikely that we will see any substantial movement in the way that we travel for either our daily mobility 'needs' or those needs met by air travel and the desire to shrink space and consume more and more places. Accordingly, we may wish to ask what kind of future there might be, as Dennis and Urry (2009) have done, in an era 'after the car'.

Placing the car

There is no end of authors who would like to propose apocalyptic outcomes for a future that sees the car relegated to nothing more than a museum piece, either as a result of resource scarcity or the sudden and desperate attempts to reduce carbon emissions (Dennis and Urry, 2009; Heinberg, 2004, 2007; Kunstler, 2005). Many of these accounts are well-researched and, if theses like that of peak oil are proved right, could well emerge. However, rather than pursue this route of future prediction, we want to end by providing a manifesto for placing the car – an item of personal mobility that is unlikely to be absent any time soon – into a place-based context that sits alongside the benefits of places that are well-planned, low mobility and have a mixed economy of transport options.

First and foremost, we do not see that there is one unifying view of what a sustainable transport system and its placement in a particular context might look like. As we have argued earlier in this chapter, what constitutes a low mobility, low carbon and high well-being experience in a particular place is a democratic issue for those residing locally which cannot be imposed from above. Second, there are some broad principles that might be applied to placing well-being and the importance of the local at the centre of decisions to rebalance transport options.

At the broadest scale, this does mean that national governments need to contend with their obsession for growth at any cost (Jackson, 2005). Dobson (2010) has argued for a radical rethink of the economic growth model applied by neo-liberal states, within the context of environmental citizenship. In essence, he argues that it is not possible to make meaningful changes to lifestyles and promote meaningful modifications in practices if the macro-economic signals point towards more travel as a signifier of economic success. This means a radical appraisal of how our economy is structured and the extent to which global modes of production, distribution and consumption (of both products and services like tourism) can be relocalised is necessary.

Such a project is long-term, fraught with political difficulties and rests on us challenging some of the most embedded beliefs about economic prosperity that have emerged since the mid nineteenth century and the growth of capitalism (Jackson, 2005). Accordingly, making an impact in this context is likely to be slow, minimal and frustrating. Yet it is a duty of academics and all those with an interest in the future to tackle head-on the underpinning factors that lead to our insatiable desire for consuming things and places.

If, however, we wanted to ground progress in something more tangible, then we can look at the places we live and ask the basic but essential question that we have been trying to pose throughout this book, which is why we like to move so much and why the places we call home seem to be environments that do not promote dwelling, well-being and satisfaction. In part this is a question of personal reflection; what is it about ourselves, our identities and our consumption habits that makes us so addicted to mobility? From the perspective of this book, it is about an engagement with what makes place meaningful: good architectural design, a sense and celebration of heritage, high-quality open spaces, a vibrant retail and food economy, and a design that promotes both pedestrianisation and dwelling.

What this would look like in a particular place depends on the topography, heritage, local building materials, climate and even the food tastes of local people. Perhaps one of the most important factors is what Kunstler (1994, 1998) refers to as human scale. That is, buildings and other structures should be commensurate with the ability of humans to relate to them as places of dwelling. Indeed, from a transport perspective, the ways in which we design our roads and fit them with other mobility technologies is critical. Of course, the private car has a place, but for too long it has dominated the built environment such that it is the human, not the machine, who has had to become marginalised in the landscape. Rather, the car is something that needs to work for us as a piece of mobility technology just as bicycles, trains and trams do. Practically speaking, this means designing streets that give preference to pavement space for pedestrians, cafes, restaurants, street performers and anything that wants to use the street as a space for activity rather than pure motorised mobility (Banister, 2008).

Such changes also mean amendments in the ways we plan places that will be the homes for new communities. At present, we follow a model of development that is dominated by suburban sprawl and the creation of housing estates with only few nods to local architectural traditions, heritage, local building materials and the essentials of what makes living in a place worthwhile – a sense of belonging, diversity in environments and a centre to commercial and cultural life.

Little wonder then that people are so intent not only to travel for work but also to get away from such places to experience the pleasures of other environments. As we have demonstrated in this book, the demand to engage in more and more tourism activities has generated a boom in flying for short breaks – everything from golfing weekends to stag and hen parties in places that are thousands of miles from home. What we argue here is that this obsession with getting away is partly related to what we are trying to get away from – places that are not what they could be. We are not arguing for a parochial abstinence from travel to foreign places, but rather a rebalancing of exotic travel with the pleasures of being at home and of working to make the places we live in those that promote well-being and quality of life.

Finally, we return to the issue of behavioural change, one that has dominated the political landscape and which for many holds the key to reducing the reliance on motor cars and flying. We have argued in this book that individualistic approaches to behavioural change are a limited response to the challenge of engaging publics in a process of change. Indeed, we have argued that current approaches need to be seen in political context as part of a neo-liberal architecture for promoting greater choice and devolving the responsibility for environmental challenges down to individual consumers (Barr and Prillwitz, 2014; Whitehead *et al.*, 2011). As such, we are limited in the real choices we can exercise because of the choice architectures in which we are placed. Accordingly, the choices we have are fundamentally limited by the political and economic system that surrounds our everyday and tourist lives. Making a choice, for example, to reduce car use for work or take the train for a holiday is not necessarily something that can be attained without considerable sacrifices in terms of time, money or both. Indeed, even if people did want to exercise their right to such choice, in many cases it is simply not possible. We have, in effect, created an economic and political system that rewards individual choices above those of the collective and so it is no surprise that individuals choose to do what they deem best for their own well-being and satisfaction – which is to consume and to travel. To try to work against this with messages of reduced consumption, as Peattie and Peattie (2009) have argued, is largely futile, because it works against the signals that individuals receive as a matter of course in their everyday lives.

Perhaps the biggest challenge, therefore, is not trying to get people to act in ways that are fundamentally contradictory to the system in which they reside, but to challenge that system. This is somewhat of a grand statement, but it is unlikely that we will be able to meet the changes needed to reduce carbon emissions if we do not begin to change the incentives that lie behind current practices. This means fundamental change in how we practice politics, the kinds of economic policies we pursue and understanding the impacts these have on travel practices. As a start, we need to consider the places in which we dwell and the lives we live in those places. It is our very strong contention that if we do not grapple with some of the challenges of what we call home, we are unlikely to ever lose the insatiable desire to travel more and more. In essence, the question we leave you with at the end of this book is one about who we are and what we value, and what makes life worth living. For most, it is not likely to be money or possessions or incessant travel, but it is likely to be about belonging and having meaningful, loving and sustained relationships.

Bibliography

Abbott, C. 1991. Urban design in Portland, Oregon, as policy and process, 1960–1989. *Planning Perspectives*, 6, 1–18.

Abbott, C. 1994. The Oregon planning style, in *Planning the Oregon Way*, edited by C. Abbott, D. A. Howe and S. Alder. Corvallis, OR: Oregon State University Press. 205–226.

Abbott, C. 1997. The Portland region: where city and suburbs talk to each other—and often agree. *Housing Policy Debate*, 8, 11–51.

ABTA Consumer Survey. 2014. *The consumer holiday trends report*. London: ABTA.

Acker, V., Goodwin, P. and Witlox, F. 2016. Key research themes on travel behaviour lifestyle and sustainable urban mobility. *International Journal of Sustainable Transportation*, 10 (1), 25–35.

Adey, P. 2006. If mobility is everything then it is nothing: towards a relational politics of (im)mobilities. *Mobilities*, 1, 75–95.

Adey, P. 2008. Architectural geographies of the airport balcony: mobility, sensation and the theatre of flight. *Geografiska Annaler: Series B, Human Geography*, 90, 29–47.

Adey, P. 2010. *Mobility*. London: Routledge.

Adey, P., Budd, L. and Hubbard, P. 2007. Flying lessons: exploring the social and cultural geographies of global air travel. *Progress in Human Geography*, 31 (6), 773–791.

Adger, W. N. 2000. Social and ecological resilience: are they related? *Progress in Human Geography*, 24, 347–364.

Aiken, G. 2012. Community transitions to low carbon futures in the Transition Towns Network (TTN). *Geography Compass*, 6 (2), 89–99.

Ajzen, I. 1991. The theory of planned behavior. *Organizational Behavior and Human Decision Processes*, 50, 179–211.

Albino, V., Berardi, U. and Dangelico, R. M. 2015. Smart cities: definitions, dimensions, performance, and initiatives. *Journal of Urban Technology*, 22, 3–21.

Anable, J. 2005. 'Complacent car addicts' or 'aspiring environmentalists'? Identifying travel behaviour segments using attitude theory. *Transport Policy*, 12, 65–78.

Anderson, K., Bows, A. and Footitt, A. 2007. *Aviation in a low carbon EU*. A research report by the Tyndall Centre, University of Manchester. Manchester: Friends of the Earth.

Andreasen, A. R. (ed.). 2006. *Social marketing in the 21st century*. London: Sage.

Archer, J., Sandul, P. J. P. and Solomonson, K. 2015. Making, performing, living suburbia, in *Making Suburbia: New Histories of Everyday America*, edited by J. Archer, P. J. P. Sandul and K. Solomonson. Minneapolis, MN: University of Minnesota Press.

Arentze, T. and Timmermans, H. 2008. Social networks, social interactions, and activity-travel behavior: a framework for microsimulation. *Environment and Planning B: Planning and Design*, 35, 1012–1027.

Ashutosh, I. and Mountz, A. 2012. The geopolitics of migrant mobility: tracing state relations through refugee claims, boats, and discourses. *Geopolitics*, 17, 335–354.

Associated Rediffusion. 1962. *London profiles, people and holidays no 2*. British Library YD.2009.a.4589.

Axhausen, K. 2008. Social networks, mobility biographies, and travel: survey challenges. *Environment and Planning B: Planning and Design*, 35, 981–996.

Bailey, I., Hopkins, R. and Wilson, G. 2010. Some things old, some things new: the spatial representations and politics of change of the peak oil relocalisation movement. *Geoforum*, 41, 595–605.

Baldassare, M. 1986. *Trouble in paradise: the suburban transformation in America*. New York: Columbia University Press.

Baldassare, M. 1992. Suburban communities. *Annual Review of Sociology*, 18, 475–494.

Bamberg, S. and Rölle, D. 2003. Determinants of people's acceptability of pricing measures: replication and extension of a causal model, in *Acceptability of Transport Pricing Strategies*, edited by J. Schade and B. Schlag. London: Pergamon Press. 235–248.

Bamberg, S. and Schmidt, P. 1998. Changing travel-mode choice as rational choice: results from a longitudinal intervention study. *Rationality and Society*, 10, 223–252.

Bamberg, S., Fujii, S., Friman, M. and Gärling, T. 2011. Behaviour theory and soft transport policy measures. *Transport Policy*, 18, 228–235.

Banister, D. 2002. *Transport planning*. 2nd edition. London and New York: SPON Press.

Banister, D. 2008. The sustainable mobility paradigm. *Transport Policy*, 15, 73–80.

Baritz, L. 1989. *The good life: the meaning of success for the American middle class*. New York: Harper and Row.

Barr, S. 2004. Are we all environmentalists now? Rhetoric and reality in environmental decision-making. Geoforum, 35, 231–249.

Barr, S. 2011. Climate forums: virtual discourses on climate change and the sustainable lifestyle. *Area*, 43 (1), 14–22.

Barr, S. 2015. Beyond behaviour change: social practice theory and the search for sustainable mobility, in *Putting Sustainability into Practice: Applications and Advances in Research on Sustainable Consumption*, edited by E. Huddart Kennedy, M. J. Cohen and N. T. Krogman. Cheltenham: Edward Elgar. 91–108.

Barr, S. 2017. Knowledge, expertise and engagement. *Environmental Values*, 26 (2), 125–130.

Barr, S. and Devine-Wright, P. 2012. Resilient communities: sustainabilities in transition. *Local Environment*, 17, 525–532.

Barr, S. and Gilg, A. W. 2006. Sustainable lifestyles: framing environmental action in and around the home. Geoforum, 37, 906–920.

Barr, S. and Pollard, J. 2017. Geographies of transition: narrating environmental activism in an age of climate change and 'peak oil'. Environment and Planning A, 49 (1), 47–64.

Barr, S. and Prillwitz, J. 2011. Green travelers? Exploring the spatial context of sustainable mobility styles. Applied Geography, *32*, 798–809.

Barr, S. and Prillwitz, J. 2014. A smarter choice? Exploring the behaviour change agenda for environmentally sustainable mobility. Environment and Planning C, 32, 1–19.

Barr, S., Gilg, A. W. and Ford, N. J. 2001. A conceptual framework for understanding and analysing attitudes towards household waste management. *Environment and Planning A*, 33, 2025–2048.

Barr, S., Gilg, A. W. and Shaw, G. 2011a. 'Helping people make better choices': exploring the behaviour change agenda for environmental sustainability. *Applied Geography*, 31, 712–720.

Barr, S., Shaw, G. and Coles, T. E. 2011b. Times for (un)sustainability? Challenges and opportunities for developing behaviour change policy. A case-study of consumers at home and away. *Global Environmental Change*, 21, 1234–1244.

Barr, S., Shaw, G., Coles, T. E. and Prillwitz, J. 2010. 'A holiday is a holiday': practicing sustainability, home and away. *Journal of Transport Geography*, 18, 474–481.

Bassett, K. 1993. British port privatization and its impact on the port of Bristol. *Journal of Transport Geography*, 1, 255–267.

Bath Spa University. 2017. *Greenwood's map: an outline history*. [Online]. Available at: http://users.bathspa.ac.uk/greenwood/lhistory.html [accessed: 21st June 2017].

Bauman, Z. 1998. *Globalization: the human consequences*. New York: Columbia University Press.

BBC. 2012. *PM urges 'garden suburb principles' in future developments*. [Online]. Available at: www.bbc.co.uk/news/uk-politics-17434001 [accessed: 14th August 2014].

BBC. 2013. *Detroit: from American dream to American nightmare*. [Online]. Available at: www.bbc.co.uk/news/24633916 [accessed: 14th August 2014].

BBC. 2014. *London on film: sprawl, sex and garden gnomes*. [Online]. Available at: www.bbc.co.uk/programmes/p00vh4cy [accessed: 14th August 2014].

BBC. 2015. *London cyclists in 23,000 accidents over five years*. [Online]. Available at: www.bbc.co.uk/news/uk-england-london-31612253 [accessed: 16th August 2016].

Becken, S. 2007. Tourists' perception of international air travel's impact on the global climate and potential climate change policies. *Journal of Sustainable Tourism*, 15, 351–368.

Becker, U., Gerike, R. and Völlings, A. 1999. *Gesellschaftliche Ziele von und für Verkehr (Social Goals of and for Transport)*. Book 1 (Dresden Institute for Transport and Environment). Dresden: DIVU.

Beckmann, J. 2001. Automobility – a social problem and theoretical concept. *Environment and Planning D*, 19 (5), 593–607.

Begg, D. and Gray, D. 2004. Transport policy and vehicle emission objectives in the UK: is the marriage between transport and environmental policy over? *Environmental Science & Policy*, 7, 155–163.

Berghoff, H.m Korte, R., Schneider, R. and Harvie, C. (eds.). 2002. *The making of modern tourism*. London: Palgrave.

Bickerstaff, K., Tolley, R. and Walker, G. 2002. Transport planning and participation: the rhetoric and realities of public involvement. *Journal of Transport Geography*, 10, 61–73.

Bissell, D. 2009a. Visualising everyday geographies: practices of vision through travel-time. *Transactions of the Institute of British Geographers*, 34, 42–60.

Bissell, D. 2009b. Conceptualising differently-mobile passengers: geographies of everyday encumbrance in the railway station. *Social and Cultural Geography*, 10, 173–195.

Black, W. R. 2000. Socio-economic barriers to sustainable transport. *Journal of Transport Geography*, 8, 141–147.

Blake, J. 1999. Overcoming the 'value-action gap' in environmental policy: tensions between national policy and local experience. *Local Environment*, 4, 257–278.

Blewitt, J. 2006. *The ecology of learning: sustainability, life-long learning and everyday life*. London: Earthscan.

Blunt, A. 2005. Cultural geography: cultural geographies of home. *Progress in Human Geography*, 29 (4), 505–515.

Blunt, A. and Dowling, R. 2006. *Home* (Key ideas in geography). Abingdon: Routledge.

Böhler, S., Grishkat, S., Haustein, S. and Hunecke, M. 2006. Encouraging environmentally sustainable holiday travel. *Transportation Research Part A: Policy and Practice*, 40 (8), 652.

Bonham, J. 2006. Transport: disciplining the body that travels, in *Against Automobility*, edited by S. Bohm, C. Jones, C. Land and M. Paterson. Oxford: Blackwell. 57–74.

Booker, F. 1977. *The great western railway: a new history*. Newton Abbot: David & Charles Publishers.

Booth, C. and Richardson, T. 2001. Placing the public in integrated transport planning, *Transport Policy*, 8, 141–149.

Bourdieu, P. 1984. *Distinction: a social critique of the judgement of taste*. London: Routledge.

Brinkley, D. 2003. *Wheels for the world: Henry Ford, his company and a century of progress*. London: Viking Press.

British National Travel Survey. 1976. [Online]. Available at: www.ukdataservice.ac.uk [accessed: 13th November 2016].

British Railways Board. 1963. *The re-shaping of Britain's railways*. London: HM Stationery Office.

Bryman, A. 2011. *Social research methods*. 4th edition. Oxford: Oxford University Press.

Budd, L. and Adey, P. 2009. The software-simulated airworld: anticipatory code and affective aeromobilities. *Environment and Planning A*, 41, 1366–1385.

Buehler, R. and Pucher, J. 2011. Sustainable transport in Freiburg: lessons from Germany's environmental capital. *International Journal of Sustainable Transportation*, 5, 43–70.

Bulkeley, H. and Castán Broto, V. 2013. Government by experiment? Global cities and the governing of climate change. *Transactions of the Institute of British Geographers*, 38 (3), 361–375.

Bull, M. 2004. Automobility and the power of sound. *Theory, Culture & Society*, 21 (4–5), 243–259.

Burgess, J., Harrison, C. M. and Filius, P. 1998. Environmental communication and the cultural politics of environmental citizenship. *Environment and Planning A*, 30, 1445–1460.

Burkart, A. J. and Medlik, S. 1981. *Tourism: past, present and future*. London: Heinemann.

Burns, P. and Bibbings, L. 2009. The end of tourism? Climate change and societal changes. *Twenty-First Century Society*, 4 (1), 31–51.

Cabinet Office. 2004. *Personal responsibility and changing behaviour*. London: Cabinet Office.

Caro, R. 1974. *The power broker: Robert Moses and the fall of New York*. New York: Vintage Books.

Carr, N. 2002. The tourism leisure behavioural continuum. *Annals of Tourism Research*, 29 (4), 972–986.

Carrasco, J. and Miller, E. 2009. The social dimension in action: a multilevel, personal networks model of social activity frequency between individuals. *Transportation Research Part A*, 43 (1), 90–104.

Carvalho, L. 2015. Smart cities from scratch? A socio-technical perspective. *Cambridge Journal of Regions, Economy and Society*, 8 (1), 43–60.

Cavada, M., Hunt, D. and Rogers, C. 2014. Smart cities: contradicting definitions and unclear measures'. *Proceedings of the 4th World Sustainability Forum 2014*. [Online]. Available at: www.researchgate.net/profile/Marianna_Cavada/publication/267764451_Smart_Cities_Contradicting_Definitions_and_Unclear_Measures/links/545a28160cf-2bccc4913249b.pdf [accessed: 21st June 2017].

Central Office of Information (COI). 1948. Charley in *New Town*. London: Central Office of Information for the Ministry of Town and Country Planning.

Chapman, L. 2007. Transport and climate change: a review. *Journal of Transport Geography*, 15, 354–367.

City of Portland. 2016. CC2035 proposed draft. Portland, OR: City of Portland.

Clapson, M. 2003. Suburban century: social change and urban growth in England and the United States. Oxford: Berg.

Clarke, J., Newman, J., Smith, N., Vidler, E. and Westmarland, L. 2007. *Creating citizen-consumers: changing publics and changing public services*. London: Sage.

Cloke, P. 1993. On 'problems and solutions'. The reproduction of problems for rural communities in Britain during the 1980s. *Journal of Rural Studies*, 9, 113–121.

Cloke, P. J., Philo, C. and Sadler, D. 2007. *Approaching human geography*. London: Sage.

Cohen, E. 1972. Toward a sociology of international tourism. *Annals of Tourism Research*, 39 (2), 164–182.

Cohen, S., Higham, J. E. S. and Cavaliere, C. T. 2011. Binge flying: behavioural addiction and climate change. *Annals of Tourism Research*, 38, 1070–1089.

Conley, J. and McLaren, A. T. (eds.). 2009. *Car troubles: critical studies of automobility and auto-mobility*. Aldershot: Ashgate.

Connell, J. and Page, S. 2008. Exploring the spatial patterns of car-based tourist travel in Lock Lomond and Trossachs National Park. *Tourism Management*, 29 (3), 561–580.

Connelly, J., Smith, G., Benson, D. and Saunders, C. 2012. *Politics and the environment*. Abingdon: Routledge.

Cresswell, T. 2010. Towards a politics of mobility. *Environment and Planning D: Society and Space*, 28, 17–31.

Cresswell, T. 2011. Mobilities: catching up. *Progress in Human Geography*, 35, 550–558.

Cresswell, T. and Merriman, P. 2011. Introduction: geographies of mobilities: practices, spaces, subjects, in *Geographies of Mobilities*, edited by T. Cresswell. Aldershot: Ashgate. 1–18.

Crompton, T. and Thøgersen, J. 2009. *Simple and painless? The limitations of spillover in environmental campaigning*. [Online]. Available at: http://assets.wwf.org.uk/downloads/simple_painless_report.pdf [accessed: 20th August 2016].

Cullingworth, B. and Nadin, V. 2006. *Town and country planning in the UK*. 14th edition. London: Routledge.

Dagger, R. 2003. Stopping sprawl for the good of all. The case for civic environmentalism. *Journal of Social Philosophy*, 34, 28–43.

Dallen, J. 2007. Sustainable transport, market segmentation and tourism: the Looe Valley branch line railway, Cornwall, UK. *Journal of Sustainable Tourism*, 15, 180–199.

Dargay, J. and Hanly, M. 2007. Volatility of car ownership, commuting mode and time in the UK. *Transportation Research Part A: Policy and Practice*, 41, 934–948.

Davison, L. J. and Ryley, T. J. 2010. Tourism destination preferences of low-cost airline users in the East Midlands. *Journal of Transport Geography*, 18 (3), 458–465.

Deffeyes, K. S. 2001. *Hubbert's peak: the impending world oil shortage*. Princeton, NJ: Princeton University Press.

De Groot, J. and Steg, L. 2007. General beliefs and the theory of planned behavior: the role of environmental concerns in the TPB. *Journal of Applied Social Psychology*, 37, 1817–1836.

De Lyser, D. 2011. Flying, feminism and mobilities–crusading for aviation in the 1920s, in *Geographies of Mobilities*, edited by T. Cresswell. Aldershot: Ashgate. 83–98.

Dennis, K. and Urry, J. 2009. *After the car*. Cambridge: Polity Press.

Department for Business, Energy and Industrial Strategy. 2016. *Building on success and learning from experience: an independent review of the research excellence framework*. London: Department for Business, Energy and Industrial Strategy.

Department for Communities and Local Government (DCLG). 2012. *National planning policy framework.* London: Department for Communities and Local Government.

Department for Communities and Local Government (DCLG). 2014. *Locally-led garden cities.* London: Department for Communities and Local Government.

Department for Environment, Food and Rural Affairs (DEFRA). 2005. *Securing the future: UK government sustainable development strategy.* London: The Stationery Office.

Department for Environment, Food and Rural Affairs (DEFRA). 2008a. *Climate change Bill.* London: HMSO.

Department for Environment, Food and Rural Affairs (DEFRA). 2008b. *A framework for pro-environmental behaviours.* London: Department of the Environment Food and Rural Affairs.

Department for Environment, Food and Rural Affairs (DEFRA). 2011. *Sustainable lifestyles framework.* London: Department of the Environment Food and Rural Affairs.

Department for Transport (DfT). 1996. *National cycling strategy.* London: HMSO.

Department for Transport (DfT). 2004a. *Smarter choices: changing the way we travel. Final report of the research project: the influence of 'soft' factor interventions on travel demand.* London: Department for Transport.

Department for Transport (DfT). 2004b. *Full guidance on local transport Plans.* 2nd edition. London: HMSO. December 2004.

Department for Transport (DfT). 2008. *Public experiences of and attitudes to air travel,* quoted in Hares, A., Dickinson, J. and Wilkes, K. 2010.

Department for Transport (DfT). 2009. *Transport trends 2009.* London: HMSO.

Department for Transport (DfT). 2011. *Climate change and transport choices. Segmentation model: a framework for reducing CO_2 emissions from personal travel.* London: Department for Transport.

Department for Transport (DfT). 2014. *National travel survey 2013.* London: Department for Transport.

Department for Transport (DfT). 2016a. *Local area walking and cycling statistics: England 2014/15 statistical release July 2016.* London: Department for Transport.

Department for Transport (DfT). 2016b. *National travel survey 2015.* London: Department for Transport.

Department for Transport (DfT). 2016c. *Single departmental plan: 2015 to 2020.* London: Department for Transport.

Department of the Environment (DoE). 1994. *Sustainable development, the UK strategy.* London: HMSO.

Department of the Environment, Transport and the Regions (DETR). 1998. *A new deal for transport: better for everyone, the government's white paper on the future of transport.* London: HMSO.

Department of the Environment, Transport and the Regions (DETR). 1999. *The UK strategy for sustainable development. A better quality of life. A strategy for sustainable development for the United Kingdom.* London: HMSO.

Department of the Environment, Transport and the Regions (DETR). 2000a. *Guidance on the methodology for multi-modal studies.* London: HMSO.

Department of the Environment, Transport and the Regions (DETR). 2000b. *Transport 2010. The 10 year plan.* London: HMSO.

Department of the Environment, Transport and the Regions (DETR). 2011. *PPG13. Transport.* London: HMSO.

Dicken, P. 2007. *Global shift: mapping the changing contours of the world economy.* London: Sage.

Dickinson, J. E. and Dickinson, J. A. 2006. Local transport and social representations: challenging the assumptions for sustainable tourism. *Journal of Sustainable Tourism*, 14, 192–208.

Dickinson, J. E. and Lumsdon, L. M. 2010. *Slow travel and tourism*. London: Earthscan.

Dickinson, J. E., Lumsdon, L. M. and Rubbins, D. 2011. Slow travel issues for tourism and climate change. *Journal of Sustainable Tourism*, 19 (3), 281–300.

Dickinson, J. E., Calver, S., Walters, K. and Wilkes, K. 2004. Journeys to heritage attractions in the UK: a case study of national trust property visitors in the South West. *Journal of Transport*, 12 (2), 103–113.

Dobson, A. 2010. *Environmental citizenship and pro-environmental behaviour: rapid research and evidence review*. London: Sustainable Development Research Network.

Docherty, I. and Shaw, J. (eds.). 2003. A *new deal for transport?* RGS-IBG Book Series. Oxford: Blackwell.

Dowling, R. and Mee, K. 2007. Home and homemaking in contemporary Australia. *Housing, Theory and Society*, 24 (3), 161–165.

Dresner, S. 2002. *The principles of sustainability*. London: Earthscan Publications Ltd.

Dugundji, E., Páez, A. and Arentze, T. 2008. Social networks, choices, mobility, and travel. *Environment and Planning B: Planning and Design*, 35, 956–960.

Dugundji, E., Páez, A., Arentze, T. A., Walker, J., Carrasco, J. A., Marchal, F. and Nakanishi, H. 2011. Transportation and social interactions. *Transportation Research Part A: Policy and Practice*, 45, 239–247.

Dunckel Graglia, A. 2016. Finding mobility: women negotiating fear and violence in Mexico City's public transit system. *Gender, Place & Culture*, 23, 624–640.

Dundon-Smith, D. M. and Gibb, R. A. 1994. The channel tunnel and regional economic development. *Journal of Transport Geography*, 2, 178–189.

Edensor, T. 2007. Mundane mobilities, performances and spaces of tourism. *Social and Cultural Geography*, 8 (2), 199–215.

Edensor, T. 2011. Commuter mobility, rhythm and commuting, in *Geographies of Mobilities*, edited by T. Cresswell. Aldershot: Ashgate. 189–204.

Edensor, T. (ed.). 2012. *Geographies of rhythm: nature, place, mobilities and bodies*. Aldershot: Ashgate.

Eijelaar, E., Peeters, P. and Piket, P. 2010. *European cycle tourism for sustainable regional development*. Breda: Centre for Sustainable Tourism and Transport, NHTV Breda University of Applied Sciences.

Elliot, R. 1994. Addictive consumption: function and fragmentation in postmodernism. *Journal of Consumer Policy*, 17 (2), 159–179.

English Tourism Council. 2002. *e-Tourism in England*. London: ETC.

Ettema, D. and Schwaren, T. 2012. A rational approach to analysing leisure travel. *Journal of Transport Geography*, 24, 173–181.

Etzioni, A. 1993. *The spirit of community: rights, responsibilities, and the communitarian agenda*. London: Fontana/Harper-Collins.

Etzioni, A. 1995. Introduction, in *New Communitarian Thinking: Persons, Virtues, Institutions, and Communities*, edited by A. Etzioni. London: University of Virginia Press. 1–15.

Euromonitor International. 2007. *WTM Global trends report 2007*. London: Euromonitor.

Faulk, P., Richie, B. and Fluker, M. 2006. *Cycle tourism in Australia: an investigation into its size and scope*. Gold Coast, QLD: Sustainable Tourism and CRC.

Featherstone, M. 2004. Automobilities: an introduction. *Theory, Culture and Society*, 21 (4), 1–24.

Fincham, B., McGuiness, M. and Murray, L. 2010. *Mobile methodologies*. London: Palgrave Macmillan.

Fishbein, M. and Ajzen, I. 1975. *Belief, attitude, intention and behavior: an introduction to theory and research*. Reading, MA: Addison-Wesley.

FitzRoy, F. and Smith, I. 1998. Public transport demand in Freiburg: why did patronage double in a decade? *Transport Policy*, 5, 163–173.

Forester, J. 1999. *The deliberative practitioner: encouraging participatory planning processes*. Boston, MA: MIT Press.

Fotis, J., Buhalis, D. and Rossides, N. 2012. Social media and impact during the holiday travel planning process, in *Information and Communication in Tourism*, edited by M. Fuchs, F. Ricci and L. Cantoni. Vienna: Springer. 13–24.

Francis, G., Dennis, N., Ison, S. and Humphreys, I. 2007. The transferability of the low-cost model to long haul airline operations. *Tourism Management*, 28 (2), 391–398.

Francis, G., Humpheys, I., Ison, S. and Aicken, M. 2006. What next for the low-cost airlines? A spatial and temporal comparative study. *Journal of Transport Geography*, 14 (2), 83–94.

Frank, R. H. 1990. Rethinking rational choice, in *Beyond the Marketplace: Rethinking Economy and Society*, edited by R. Friedland and A. F. Robertson. Piscataway, NJ: Aldine Transaction. 53–87.

Franks, T. R. 1996. Managing sustainable development: definitions, paradigms and dimensions. *Sustainable Development*, 4 (2), 53–60.

Freiria, S., Ribeiro, B. and Tavares, A. O. 2015. Understanding road network dynamics: link-based topological patterns. *Journal of Transport Geography*, 46, 55–66.

French, J., Blair-Stevens, C., McVey, D. and Merritt, R. 2009. *Social marketing and public health: theory and practice*. Oxford: Oxford University Press.

French, J., Blair-Stevens, C., McVey, D. and Merritt, R. (eds.). 2010. *Social marketing and public health: theory and practice*. Oxford: Oxford University Press.

Freudendal-Pedersen, M. 2009. *Mobility in daily life*. Aldershot: Ashgate.

Fujii, S., Gärling, T. and Kitamura, R. 2001. Changes in drivers' perceptions and use of public transport during a freeway closure. *Environment and Behavior*, 33, 796–808.

Fullager, S., Wilson, E. and Markwell, K. 2012. Starting slow: thinking slow mobilities and experiences, in *Slow Tourism: Experiences and Mobilities*, edited by S. Fullager, E. Wilson and K. Markwell. Bristol: Channel View. 1–10.

Gabrys, J. 2014. Programming environments: environmentality and citizen sensing in the smart city. *Environment and Planning D: Society and Space*, 32 (1), 30–48.

Gans, H. J. 1967. *The levittowners: ways of life and politics in a new suburban community*. London: Penguin.

Gardner, N. 2009. A manifesto for slow travel. *Hidden Europe Magazine*, 25, 10–14.

Gärling, T. and Axhausen, K. W. 2003. Introduction: habitual travel choice. *Transportation*, 30, 1–11.

Gaytours Brochure. 1964. *Why travel with gaytours*. British Library LD.31.a.724.

Gerike, R. 2007. *How to make sustainable transportation a reality? The development of three constitutive task fields for transportation*. Munich: Oekom.

Geurs, K. T. and van Wee, B. 2004. Accessibility evaluation of land-use and transport strategies: review and research directions. *Journal of Transport Geography*, 12, 127–140.

Giddens, A. 1991. *Modernity and self-identity: self and society in the late modern age*. London: Polity Press.

Gilg, A. W. 1996. *Countryside planning*. 2nd edition. London: Sage.

Gilg, A. W. 2005. *Planning in Britain: understanding and evaluating the post-war system.* London: Sage.

Glaser, E. L. and Shapiro, J. 2001. *Is there a new urbanism? The growth of US cities in the 1990's.* Cambridge, MA: National Bureau of Economic Research. Working Paper No. 8357.

Glasmeier, A. and Christopherson, S. 2015. Thinking about smart cities. *Cambridge Journal of Regions, Economy and Society*, 9, 3–12.

Glasmeier, A. and Nebiolo, M. 2016. Thinking about smart cities: the travels of a policy idea that promotes a great deal, but so far has delivered modest results. *Sustainability*, 8 (11), 1122. doi: 10.3390/su8111122.

Goetz, A., Vowles, T. and Tierney, S. 2009. Bridging the qualitative-quantitative divide in transport geography. *The Professional Geographer*, 61, 323–335.

Goodwin, P. 1999. Transformation of transport policy in Great Britain. *Transportation Research, Part A*, 33, 655–669.

Gorr, H. 1997. *Die Logik der individuellen Verkehrsmittelwahl Theorie und Entscheidungsverhalten im Personenverkehr* (The logic of individual travel mode choice. Theory and decision behaviour in passenger traffic). Gießen: Focus.

Gorringe, T. 2002. *The theology of the built environment: justice, empowerment, redemption.* Cambridge: Cambridge University Press.

Gössling, S. and Peeters, P. 2007. 'It does not harm the environment!' An analysis of industry discourses on tourism, air travel and the environment. *Journal of Sustainable Tourism*, 15, 402–417.

Götz, K., Loose, W., Schmied, M. and Schubert, S. 2003. *Mobility styles in leisure time.* Paper presented at the 10th International Conference on Travel Behaviour Research, Lucerne, 10th to 15th August 2003.

Graf, L. 2005. Incompatibilities of the low-cost and network carrier business models within the same airline grouping. *Journal of Air Transport Management*, 11 (5), 313–327.

Graham, B. and Shaw, J. 2008. Low-cost airlines in Europe: reconciling liberalization and sustainability. *Geoforum*, 39, 1439–1451.

Graham, B. and Vowles, J. M. 2006. Carries within carriers: a strategic response to low-cost airline competition. *Transport Reviews*, 26 (1), 105–126.

Green, D. P. and Shapiro, I. 1994. *Pathologies of rational choice theory: a critique of applications in political science.* Newhaven, CT: Yale University Press.

Grindrod, J. 2013. *Concretopia: a journey around the rebuilding of post-war Britain.* Brecon: Old Street Publishing.

The Guardian. 2015. *Funding cuts to local bus services leave people isolated – labour.* 4th January 2015. [Online]. Available at: www.theguardian.com/uk-news/2015/jan/04/funding-cuts-local-bus-services-people-isolated. [accessed: 27th August 2016].

Hägerstrand, T. 1970. What about people in regional science? *Regional Science Association Papers*, 24, 7–21.

Halden, D. and Davison, P. 2005. *10 things you need to know about accessibility planning.* [Online]. Available at: www.dhc1.co.uk/features/accessibility_planning_10_questions.html [accessed: 1st September 2016].

Hall, P. 2012. Drawing on the Garden City prescription. *Town and Country Planning*, 81, 475–478.

Hall, P., Hesse, M. and Jean-Paul, M. 2006. Re-exploring the interface between economic and transport geography. *Environment and Planning A*, 38, 1401–1408.

Halliday, S. 2013. *Underground to everywhere: London's underground railway in the life of the capital.* London: History Press.

Hammer Films. 1972. *Mutiny on the buses*. London: Hammer Films.

Hannam, K., Butler, G. and Parris, C. M. 2014. Developments and key issues in tourism mobilities. *Annals of Tourism Research*, 44 (2), 171–185.

Hannam, K., Sheller, M. and Urry, J. 2006. Editorial: mobilities, immobilities and moorings. *Mobilities*, 1, 1–22.

Hares, A., Dickinson, J. and Wilkes, K. 2010. Climate change and the air travel decisions of UK tourists. *Journal of Transport Geography*, 18 (3), 466–473

Harker Martin, B. and MacDonnell, R. 2012. Is telework effective for organizations? A meta-analysis of empirical research on perceptions of telework and organizational outcomes. *Management Research Review*, 35, 602–616.

Haworth, J. and Lewis, S. 2005. Work, leisure and well-being. *British Journal of Guidance and Counselling*, 33 (1), 67–79.

Haxeltine, A. and Seyfang, G. 2009. *Transitions for the people: theory and practice of 'transition' and 'resilience' in the UK's transition movement*. Norwich: Tyndall Centre for Climate Change Research. Working Paper, 134.

Hayden, D. 2003. *Building suburbia: green fields and urban growth, 1820–2000*. New York: Vintage Books.

Hayes, D. 2005. *Historic atlas of Vancouver and the Lower Fraser Valley*. Vancouver: Douglas & McIntyre.

Head, B. W. 2008. Wicked problems in public policy. *Public Policy*, 3, 101–118.

Headicar, P. 2009. *Transport policy and planning in Great Britain*. London: Routledge.

Heath, Y. and Gifford, R. 2002. Extending the theory of planned behavior: predicting the use of public transportation. *Journal of Applied Social Psychology*, 32, 2154–2189.

Heggie, I. G. 1978. Putting behaviour into behavioural models of travel choice. *Journal of the Operational Research Society*, 29, 541–550.

Heinberg, R. 2004. *Powerdown: options and actins for a post-carbon world*. Gabriola Island, BC: New Society Publishers.

Heinberg, R. 2007. *Peak everything: waking up to the century of decline in Earth's resources*. West Hoathly, West Sussex: Clairview Books.

Helbrecht, I. and Dirksmeier, P. (eds.). 2012. *New urbanism: life, work, and space in the New Downtown*. Aldershot: Ashgate.

Henderson, K. and Lock, K. 2012. The return of the Garden City. *Town and Country Planning*, 81, 372–375.

Hensher, D. A., Mulley, C. and Rose, J. M. 2015. Understanding the relationship between voting preferences for public transport and perceptions and preferences for bus rapid transit versus light rail. *Journal of Transport Economics and Policy*, 49, 236–260.

Hine, J. and Russell, J. 1993. Traffic barriers and pedestrian crossing behavior. *Journal of transport Geography*, 1, 230–239.

Hobson, K. 2002. Competing discourses of sustainable consumption: does the rationalisation of lifestyles' make sense? *Environmental Politics*, 11, 95–120.

Hollands, R. G. 2015. Critical interventions into the corporate smart city. *Cambridge Journal of Regions, Economy and Society*, 8 (1), 61–77.

Holloway, L. and Hubbard, P. 2001. *People and place: the extraordinary geographies of everyday life*. London: Pearson.

Hopkins, R. 2008. *The transition handbook: from oil dependency to local resilience*. Totnes: Green Books.

Horner, P. 1991. *The travel industry in Britain*. Cheltenham: Stanley Thornes.

House of Commons. 1999. *A century of change: trends in UK statistics since 1900*. House of Commons Research Paper 99/111. London: House of Commons Library.

House of Commons Papers. 2016. *Tourism: statistics and policy*. Briefing Paper 06022.

House of Lords Science and Technology Committee. 2011. *Behaviour change report*. HL Paper 179. London: The Stationery Office.

Howard, E. 1898. *To-morrow. A peaceful path to reform*. London: S. Sonnenschein & Co.

Howard, E. 1902. *Garden cities of tomorrow*. London: S. Sonnenschein & Co.

Hoyle, B. and Knowles, R. (eds.). 1992. *Modern transport geography*. Chichester: Wiley.

Huddart Kennedy, E., Cohen, M. J. and Krogman, N. T. (eds.). 2015. *Putting sustainability into practice: applications and advances in research on sustainable consumption*. Cheltenham: Edward Elgar.

Huijbens, E. and Benediktsson, K. 2007. Practising highland heterotopias: automobility in the interior of Iceland. *Mobilities*, 2 (1), 143–165.

Hull, A. 2005. Integrated transport planning in the UK: from concept to reality. *Journal of Transport Geography*, 13 (4), 318–328.

Humphreys, I. M., Ison, S. G., Francis, G. and Aldridge, K. 2005. UK airport surface access targets. *Journal of Air Transport Management*, 11 (2), 117–124.

Hunecke, M., Haustein, S., Grischkat, S. and Böhler, S. 2007. Psychological, sociodemographic, and infrastructural determinants of ecological impact caused by mobility behaviour. *Journal of Environmental Psychology*, 27, 277–292.

Huq, R. 2013. *Making sense of suburbia through popular culture*. London: Bloomsbury.

Ioannides, D. and Debbage, K. G. 1998. *The economic geography of the tourism industry: a supply-side analysis*. London: Routledge.

IPCC International Panel on Climate Change. 2001. *Climate Change 2001. The scientific basis*. [Online]. www.grida.no/climate/ipcc_tar/index.htm [accessed: 1st September 2016].

IPCC International Panel on Climate Change. 2014. *Climate Change 2014: synthesis report. Contribution of Working Groups I, II and III to the Fifth Assessment Report of the Intergovernmental Panel on Climate Change*. Core Writing Team, Pachauri, R. K. and Meyer, L. A. (eds.). [Online]. www.ipcc.ch/report/ar5/syr/ [accessed: 1st September 2016].

Iso-Ahola, S. E. 1983. Towards a social psychology of recreational travel. *Leisure Studies*, 2 (1), 45–56.

Ivy, R. L. 1993. Variations in hub service in the US domestic air transportation network. *Journal of Transport Geography*, 1, 211–218.

Jackson, A. A. 2003. The London railway suburb, 1850–1914, in *The Impact of the Railway on Society in Britain: Essays in Honour of Jack Simmons*, edited by A. K. B. Evans and J. V. Gough. Aldershot: Ashgate. 169–180.

Jackson, A. A. 2006. *London's metroland*. London: Capital Transport.

Jackson, K. T. 1985. *Crabgrass frontier: the suburbanisation of the United States*. Oxford: Oxford University Press.

Jackson, T. 2005. *Motivating sustainable consumption: a review of evidence on consumer behaviour and behavioural change*. A report to the Sustainable Development Research Network (Sustainable Development Research Network/ESRC Sustainable Technologies Programme, London).

Jain, J. 2009. The making of mundane bus journeys, in *The Culture of Alternative Mobilities: Routes Less Travelled*, edited by P. Vannini. Aldershot: Ashgate. 91–107.

Jansson, A. 2007. A sense of tourism: new media and the dialectic of encapsulation/decapsulation. *Tourist Studies*, 7 (1), 5–24.

Jeekel, H. 2013. *The car-dependent society: a European perspective*. Aldershot: Ashgate.

Jessop, B. 2002. Liberalism, neoliberalism, and urban governance: a state–theoretical perspective. *Antipode*, 34, 452–472.

Johnston, J. 2008. The citizen-consumer hybrid: ideological tensions and the case of whole foods market. *Theory and Society*, 37, 229–270.

Jones, R., Pykett, J. and Whitehead, M. 2011a. The geographies of soft paternalism in the UK: the rise of the Avuncular State and changing behaviour after neoliberalism. *Geography Compass*, 5, 50–62.

Jones, R., Pykett, J. and Whitehead, M. 2011b. Governing temptation: changing behaviour in an age of libertarian paternalism. *Progress in Human Geography*, 35, 483–501.

Jones, R., Pykett, J. and Whitehead, M. 2013. *Changing behaviours: on the rise of the psychological state*. Cheltenham: Edward Elgar.

Kamarulzaman, Y. 2007. Adoption of travel e-shopping in the UK. *International Journal of Retail and Distribution Management*, 35 (9), 703–719.

Katz, P. (ed.). 1994. *The new urbanism: toward an architecture of community*. New York: McGraw-Hill.

Keeling, D. 2007. Transportation geography: new directions on well-worn trails. *Progress in Human Geography*, 31, 217–25.

Keeling, D. 2008. Transportation geography – new regional mobilities. *Progress in Human Geography*, 32, 275–83.

Keeling, D. 2009. Transportation geography: local challenges, global contexts. *Progress in Human Geography*, 33, 516–526.

Kellett, J. R. 1969. *The impact of railways on Victorian cities*. London: Routledge & Kegan Paul.

Kenworthy, J. R. 2006. The eco-city: ten key transport and planning dimensions for sustainable city development. *Environment and Urbanization*, 18, 67–85.

Kenyon, S. and Lyons, G. 2003. The value of integrated multimodal traveller information and its potential contribution to modal change. *Transportation Research Part F*, 6, 1–21.

Kitchin, R. 2013. Big data and human geography opportunities, challenges and risks. *Dialogues in Human Geography*, 3 (3), 262–267.

Kitchin, R. 2014. The real-time city? Big data and smart urbanism. *GeoJournal*, 79 (1), 1–14.

Kitchin, R. 2015a. Making sense of smart cities: addressing present shortcomings. *Cambridge Journal of Regions, Economy and Society*, 8, 131–136.

Kitchin, R. 2015b. Positivist geography, in *Approaches to Human Geography: Philosophies, Theories, People and Practices* edited by S. C. Aitken and G. Valentine. London: Sage. 23–34.

Kitchin, R. 2015c. The promise and peril of smart cities. *Computers and law: the journal of the Society for Computers and Law*, 26 (2). Published online.

Kitchin, R. and Tate, N. 2013. *Conducting research in human geography: theory, methodology and practice*. Abingdon: Routledge.

Kleinert, M. 2009. Solitude at sea or social sailing? The constitution and perception of the cruising community, in *The Cultures of Alternative Mobilities: Routes Less Travelled*, edited by P. Vannini. Aldershot: Ashgate. 159–176.

Knights, B. 2006. In search of England: travelogue and nation between wars, in *Landscape and Englishness* edited by R. Burden and S. Kohl. Amsterdam and New York: Rodopi B.V. 165–184.

Knowles, R. D. 1993. Research agendas in transport geography for the 1990s. *Journal of Transport Geography*, 1, 3–11.

Knowles, R. D., Shaw, J. and Docherty, I. 2008. *Transport geographies: mobilities, flows and spaces*. Oxford: Blackwell.

Kowald, M., Frei, A., Hackney, K., Illenberger, J. and Axhausen, K. 2010. Collecting data on leisure travel: the link between leisure acquaintances and social interactions. *Proceedings of Social Behaviour Science*, 4 (1), 38–48.

Kunstler, J. H. 1994. *The geography of nowhere: the rise and decline of America's man-made landscape*. New York: Simon and Schuster.

Kunstler, J. H. 1998. *Home from nowhere: re-making our world for the 21st century*. New York: Simon and Schuster.

Kunstler, J. H. 2005. *The long emergency: surviving the end of oil, climate change, and other converging catastrophes of the 21st century*. New York: Simon and Schuster.

Kurz, T., Gardner, B., Verplanken, B. and Abraham, C. 2015. Habitual behaviors or patterns of practice? Explaining and changing repetitive climate-relevant actions. *Wiley Interdisciplinary Reviews: Climate Change*, 6, 113–128.

Lally, M. 2012. Lessons in neighbourhood place-shaping. *Town and Country Planning*, 81, 89–93.

Lane, S. N., Odoni, N., Landström, C., Whatmore, S. J., Ward, N. and Bradley, S. 2011. Doing flood risk science differently: an experiment in radical scientific method. *Transactions of the Institute of British Geographers*, 36, 15–36.

Langdon, P. 1994. *A better place to live: reshaping the American suburb*. Boston, MA: MIT Press.

Lanzendorf, M. and Scheiner, J. 2004. Verkehrsgenese als herausforderung für trans-disziplinarität. Stand und perspektiven (Travel genesis as a challenge for transdiscipli-narity. State of the art and prospects), in *Verkehrsgenese: Entstehung von Verkehr sowie Potenziale und Grenzen der Gestaltung einer nachhaltigen Mobilität* (Travel Genesis: Generation of Travel and Potentials and Limitations for Shaping a Sustainable Mobility), Studien zur Mobilitäts- und Verkehrsforschung (Studies of Mobility and Transportation Research) no. 5, edited by H. Dalkmann, M. Lanzendorf and J. Scheiner. Mannheim: MetaGIS. 11–38.

Larsen, J., Urry, J. and Axhausen, K. W. 2007. Networks and tourism: mobile social life. *Annals of Tourism Research*, 34, 244–262.

Larsen, J., Urry, J. and Axhausen, K. 2012. *Mobilities, networks, geographies*. Aldershot: Ashgate.

Las Vegas Review Journal. 2014. *Economic report shows Las Vegas growth, optimism are on the rise*. [Online]. Available at: www.reviewjournal.com/business/economic-report-shows-las-vegas-growth-optimism-are-rise [accessed: 14th August 2014].

Latimer, J. and Munro, R. 2006. Driving the social, in *Against Automobility*, edited by S. Bohm, C. Jones, C. Land and M. Paterson. Oxford: Blackwell. 193–207.

Laurier, E. 2011. Driving: pre-cognition and driving, in *Geographies of Mobilities*, edited by T. Cresswell. Aldershot: Ashgate. 69–82.

Laurier, E. and Dant, T. 2012. What we do whilst driving: towards the driverless car, in *Mobilities: New Perspectives on Transport and Society*, edited by M. Grecio and J. Urry. Aldershot: Ashgate. 223–244.

Laurier, E., Lorimer, H., Brown, B., Jones, O., Juhlin, O., Noble, A., Peery, M., Pica, D., Sormani, P., Strebel, I., Swan, L., Taylor, A. S., Watts, L. and Weilenmann, A. 2008. Driving and 'passengering': notes on the ordinary organization of car travel. *Mobilities*, 3, 1–23.

Leicestershire County Council. 2015. *Leicestershire's local transport plan 3*. [Online]. Available at: www.leics.gov.uk/index/highways/transport_plans_policies/ltp/current_transport_plans.htm [accessed: 7th September 2016].

Levinson, D. 2008. The orderliness hypothesis: the correlation of rail and housing development in London. *The Journal of Transport History*, 29 (1), 98–114.

Lew, A. and McKercher, B. 2008. Modelling tourist movements: a local destination analysis. *Annals of Tourism Research*, 22 (2), 403–423.

Lock, D. 2012. Garden cities – pick up the ball and run, run, run. *Town and Country Planning*, 81, 357–359.

London Transport Museum. 2014. *Online Museum*. [Online]. Available at: www.ltmcollection.org/museum/index.html [accessed: 14th August 2014].

Lorimer, H. 2011. Walking: new forms and spaces for studies of pedestrianism, in *Geographies of Mobilities*, edited by T. Cresswell. Aldershot: Ashgate. 19–33.

Lorimer, H. and Lund, K. 2003. Performing facts: finding a way over Scotland's mountains. *Sociological Review*, 51, 130–144.

Los Angeles Times. 2014. *California's population rises to 38.3 million during 2013*. [Online]. Available at: www.latimes.com/local/la-me-calif-population-20140501-story.html [accessed: 14th August 2014].

Lucas, K. and Jones, P. 2009. *The car in British society*. London: RAC Foundation for Motoring.

Lumsdon, L. M. and McGrath, P. 2011. Developing a conceptual framework for slow travel: a grounded theory approach. *Journal of Sustainable Tourism*, 19, 265–279.

Lund, H. 2003. Testing the claims of new urbanism: local access, pedestrian travel, and neighboring behaviors. *Journal of the American Planning Association*, 69, 414–429.

Luoma, M., Mikkonen, K. and Palomäki, M. 1993. The threshold gravity model and transport geography: how transport development influences the distance-decay parameter of the gravity model. *Journal of Transport Geography*, 1, 240–247.

Lupi, T. and Musterd, S. 2006. The suburban 'community question'. *Urban Studies*, 43, 803–817.

Lupton, D. 2013. *Risk*. 2nd edition. Abingdon: Routledge.

Lyons, G., Chatterjee, K., Marsden, G. and Beecroft, M. 2007. Detriments of travel demand: the future of society and lifestyles. *33rd Universities Transport Study Group Conference*. [Online]. Available at: http://eprints.uwe.ac.uk/9435 [accessed: 14th July 2014].

MacKay, K. and Vogt, C. 2012. Information technology in everyday and vacation contexts. *Annals of Tourism Research*, 39 (3), 1380–1401.

McKenzie-Mohr, D. 2000. New ways to promote proenvironmental behaviour: promoting sustainable behaviour: an introduction to community-based social marketing. *Journal of Social Issues*, 56, 543–554.

MacKinnon, D. and Derickson, K. D. 2013. From resilience to resourcefulness – a critique of resilience policy and activism. *Progress in Human Geography*, 37, 253–270.

Mahmassani, H. S. and Jou, R. C. 2000. Transferring insights into commuter behavior dynamics from laboratory experiments to field surveys. *Transportation Part A: Policy and Practice*, 34, 243–260.

Marsden, G. and Bonsall, P. 2006. Understanding the role of performance targets in transport policy. *Transport Policy*, 13 (3), 191–203.

Marsden, G., Bache I. and Kelly, C. 2012. A policy perspective on transport and climate change issues, in *Transport and Climate Change*, edited by T. J. Ryley and L. Chapman. Bingley: Emerald. 197–223.

Marsden, G., Kelly, C. and Nellthorp, J. 2009. The likely impacts of target setting and performance rewards in local transport. *Transport Policy*, 16 (2), 55–97.

Marshall, A. 2001. The challenge of sustainable transport, in *Planning for a Sustainable Future*, edited by A. Layard, S. Davoudi and S. Batty. London: Spon. 131–147.

Martinez-Garcia, E. and Raya, J. M. 2008. Length of stay for low-cost tourism. *Tourism Management*, 29 (6), 1064–1075.

Maslow, A. H. 1943. A theory of human motivation. *Psychological Review*, 50 (4), 370–396.

Mattingley, P. 2001. *Suburban landscapes: culture and politics in a New York metropolitan county* Baltimore, MD: The John Hopkins University Press.

Medearis, D. and Daseking, W. 2012. Freiburg, Germany: Germany's eco-capital, in *Green Cities of Europe*, edited by T. Beatley. Gabriola Island, BC: Island Press. 65–82.

Merriman, P. 2009. Automobility and the geographies of the car. *Geography Compass*, 3, 586–599.

Middleton, J. 2009. Stepping in time: walking, time, and space in the city. *Environment and Planning A*, 41, 1943–1961.

Middleton, J. 2010. Sense and the city: exploring the embodied geographies of urban walking. *Social & Cultural Geography*, 11, 575–596.

Miles, S. 1998. *Consumerism as a way of life*. London: Sage.

Miller, L. A. 1995. Family togetherness and the suburban ideal. *Sociological Forum*, 10, 393–418.

Miller-Boyett Productions. 1984. *Happy days*. Salisbury, CT: Miller-Boyett Productions.

Ministry of Transport. 1963. *Traffic in towns: a study of the long-term problems of traffic in urban areas*. Reports of the Steering Group and Working Group appointed by the Minister of Transport. London: HMSO.

Mintel. 2009. *Slow travel special report*. London: Mintel.

Møller, B. 2002. *Travel mode choice as habitual behaviour: a review of literature*. Aarhus: Aarhus School of Business. Working Paper 02-1.

Morton, M. V. 1935. *In search of England*. Reprinted 2002. Cambridge, MA: DE Capo Press.

Mowforth, M. and Munt, I. 1998. *Tourism and sustainability: new tourism in the third world*. 1st edition. London: Routledge.

Mowforth, M. and Munt, I. 2003. *Sustainable tourism in developing countries: poverty alleviation, participatory planning, and ethical issues*. London: Routledge.

Mowforth, M. and Munt, I. 2015. *Tourism and sustainability*. 4th edition. London: Routledge. Jersey. Chapters 1 and 2.

Munt, I. 1994. The "other" postmodern tourism: culture, travel and the new middle classes. *Theory, Culture and Society*, 11 (3), 101–125.

National Social Marketing Centre. 2016. *NSMC: leading behaviour change*. [Online]. Available at: www.thensmc.com/ [accessed: 17th August 2016].

National Trust. 2012. *National Trust comment on garden cities call*. [Online]. Available at: https://ntpressoffice.wordpress.com/2012/11/23/national-trust-comment-on-garden-cities-call/ [accessed: 14th August 2014].

Newman, P. and Kenworthy, J. 1989. *Cities and automobile dependency. An international sourcebook*. Aldershot: Gower Technical.

Nicolaides, B. M. and Wiese, A. (eds.). 2006. *The suburb reader*. London: Routledge.

Njegovan, N. 2006. Elasticities of demand for leisure air travel: a system modelling approach. *Journal of Air Transport Management*, 12, 33–39.

Nordlund, A. M. and Garvill, J. 2002. Value structures behind proenvironmental behavior. *Environment and Behavior*, 34, 740–756.

North, P. 2010. Eco-localisation as a progressive response to peak oil and climate change–a sympathetic critique. *Geoforum*, 41, 585–594.

Nutley, S. 1982. The extent of public transport decline in Wales. *Cambria*, 9, 27–48.

Nutley, S. D. 1985. Planning options for the improvement of rural accessibility: use of the time-space approach. *Regional Studies*, 19, 37–50.

Nutley, S. D. 1988. 'Unconventional modes' of transport in rural Britain: progress to 1985. *Journal of Rural Studies*, 4, 73–86.

Nutley, S. 1992. Rural areas: the accessibility problem, in *Modern Transport Geography*, edited by B. S. Hoyle and R. D. Knowles. London, Belhaven. 125–154.

O'Connell, S., 1998. *The car and British society, 1896–1939*. Manchester: Manchester University Press.

Office for National Statistics. 2013. *170 years of industrial change across England and Wales*. [Online]. Available at: www.ons.gov.uk/ons/rel/census/2011-census-analy-sis/170-years-of-industry/170-years-of-industrial-changeponent.html [accessed: 15th August 2016].

Oh, H., Assaf, A. G. and Baloglu, S. 2016. Motivations and goals of slow tourism. *Journal of Travel Research*, 55 (2), 205–219.

O'Neill, J. 2008. Happiness and the good life. *Environmental Values*, 17, 125–144.

O'Regan, M. 2011. Wandering Australia: independent travelers and slow journeys, in *Slow Tourism: Experiences and Mobilities*, edited by S. Fullager, E. Wilson and K. Markwell. Bristol: Channel View. 128–142.

Ouellette, J. A. and Wood, W. 1998. Habit and intention in everyday life: the multiple processes by which past behavior predicts future behavior. *Psychological Bulletin*, 124, 54–74.

Owens, S. 2000. Engaging the public: information and deliberation in environmental poli-cy. *Environment and Planning A*, 32, 1141–1148.

Owens, S. 2015. *Knowledge, policy and expertise*. Oxford: Oxford University Press.

Paek, H. J., Kim, S., Hove, T. and Huh, J. Y. 2014. Reduced harm or another gateway to smoking? Source, message, and information characteristics of e-cigarette videos on YouTube. *Journal of Health Communication*, 19, 545–560.

Pascal, A. 1987. The vanishing city. *Urban Studies*, 24, 597–603.

Pearce, D. G. 1987. Spatial patterns of package tourism in Europe. *Annals of Tourism Re-search*, 14 (2), 183–201.

Pearce, P. and Gretzel, H. 2012. Tourism in technology dead zones: documenting experien-tial dimensions. *International Journal of Tourism Sciences*, 12 (2), 1–20.

Peattie, K. 2012. *Social marketing: business thinking for social goals*. Presentation deliv-ered at Living with Environmental Change: Working in Partnership session at Commu-nicate Conference 2012.

Peattie, K. and Peattie, S. 2009. Social marketing: a pathway to consumption reduction? *Journal of Business Research*, 62, 260–268.

Persky, J. 1995. Retrospectives: the ethology of homo economicus. *Journal of Economic Perspectives*, 9, 221–223.

Petrick, J. and Durko, A. 2013. Family and relationship benefits of travel experience: a literature review. *Journal of Travel Research*, 52, 720–730.

Pine, J. II and Gilmore, J. H. 1999. *Experience economy: work is theatre & every business a stage*. Boston, MA: Harvard Business School Press.

Plog, S. C. 1987. Understanding psychographics in tourism research, in *Travel, Tourism and Hospitality Research: A Handbook for Managers and Researchers*, edited by J. R. Brent Ritchie and C. R. Goeldner. New York: Wiley. 203–213.

Podobnik, B. 2011. Assessing the social and environmental achievements of New Urban-ism: evidence from Portland, Oregon. *Journal of Urbanism: International Research on Placemaking and Urban Sustainability*, 4, 105–126.

Pooley, C., Turnbull, J. and Adams, M. 2005. *A mobile century? Changes in everyday mobility in Britain in the twentieth century*. Aldershot: Ashgate.

Price, J. 2000. Between subdivisions and shopping malls: signifying everyday life in the contemporary American south, in *Expanding Suburbia: Reviewing Suburban Narratives*, edited by R. Webster. Oxford: Berghahn. 125–140.

Prillwitz, J. and Barr, S. 2011. Moving towards sustainability? Mobility styles, attitudes and individual travel behaviour. *Journal of Transport Geography*, 19, 1590–1600.

Rajan, S. C. 2006. Automobility and the liberal disposition, in *Against Automobility*, edited by S. Bohm, C. Jones, C. Land and M. Paterson. Oxford: Blackwell. 113–129.

Randles, S. and Mander, P. 2009. Aviation, Consumption and the climate change debate: are you going to tell me off for flying. *Technology Analysis and Strategic Management*, 21 (1), 93–113.

Reckwitz, A. 2002. Toward a theory of social practices: a development in culturalist theorizing. *European Journal of Social Theory*, 5, 243–263.

Rickly-Boyd, J. M. 2013. 'Dirtbags': mobility, community and rock climbing as performative of identity, in *Lifestyle Mobilities: Intersections of Travel, Leisure and Migration*, edited by T. Duncan, S. C. Cohen and M. Thulemark. Aldershot: Ashgate. 51–64.

Rietveld, P. 2011. Telework and the transition to lower energy use in transport: on the relevance of rebound effects. *Environmental Innovation and Societal Transitions*, 1, 146–151.

Robinson, J. 2004. Squaring the circle? Some thoughts on the idea of sustainable development. *Ecological Economics*, 48, 369–384.

Rodrigue, J., Comtois, C. and Slack, B. 2013. *The geography of transport systems*. 3rd edition. Milton Park: Routledge.

Rojek, C. 1993. *Ways of escape; modern transformation in leisure and travel*. Houndmills: Macmillan.

Rose, N. and Miller, P. 1992. Political power and the state: problematics of government. *British Journal of Sociology*, 43, 173–205.

Ryan, C. 2010. Ways of conceptualising the tourist experience: a review of literature. *Tourism Recreation Research*, 35 (1), 37–46.

Ryan, S. and Throgmorton, J. A. 2003. Sustainable transportation and land development on the periphery: a case study of Freiburg, Germany and Chula Vista, California. *Transportation Research Part D: Transport and Environment*, 8, 37–52.

Sadik-Khan, J. and Solomonov, S. 2016. *Street fight: handbook for an urban revolution*. New York: Viking.

Sakai, T., Kawamura, K. and Hyodo, T. 2015. Locational dynamics of logistics facilities: evidence from Tokyo. *Journal of Transport Geography*, 46, 10–19.

Sawday, A. 2009. *Go slow Italy*. Bristol: Alistair Sawday Publishing.

Scammell, M. 2000. The internet and civic engagement: the age of the citizen-consumer. *Political Communication*, 17, 351–355.

Schlich, R., Schonfelder, S., Hanson, S. and Kay, A. 2007. Structures of leisure travel: temporal and spatial variability. *Transport Reviews*, 24 (2), 219–237.

Scholor, J. 2010. *The new economics of new wealth*. New York: Penguin.

Schwartz, S. H. 1977. Normative influences on altruism, in *Advances in Experimental Social Psychology*, edited by L. Berkowitz. New York: Academic Press Inc. 221–279.

Schwartz, S. H. 1992. Universals in the content and structure of values: theoretical advances and empirical test in 20 countries. *Advances in Experimental Social Psychology*, 25, 1–65.

Sen, A. 1982. *Choice, welfare and measurement*. Oxford: Blackwell.

Seyfang, G. 2005. Shopping for sustainability: can sustainable consumption promote eco-logical citizenship. *Environmental Politics*, 14 (2), 290–306.

Seyfang, G. 2006. Ecological citizenship and sustainable consumption: examining local organic food networks. *Journal of Rural Studies*, 22, 383–395.

Seyfang, G. 2010. Community action for sustainable housing: building a low-carbon fu-ture. *Energy Policy*, 38, 7624–7633.

Seyfang, G. and Haxeltine, A. 2012. Growing grassroots innovations: exploring the role of community-based initiatives in governing sustainable energy transitions. *Environment and Planning C: Government and Policy*, 30, 381–400.

Sharp, T. 1946. *Exeter Phoenix: a plan for re-building*. London: Architectural Press.

Shaw, G. 2005. Lifestyles and changes in tourism consumption: the British experience, in *Postmoderne Freizeitstile und Freizeiträume*, edited by P. Reuber and P. L. Schnell. Berlin: Erich Schmidt Verlag. 21–46.

Shaw, G. and Williams, A. 2004. *Tourism and tourism spaces*. London: Sage.

Shaw, J. and Hesse, M. 2010. Transport, geography and the 'new' mobilities. *Transactions of the Institute of British Geographers*, 35, 305–312.

Shaw, J. and Sidaway, J. D. 2010. Making links: on (re)engaging with transport and trans-port geography. *Progress in Human Geography*, 35, 502–520.

Shaw, S. and Thomas, C. 2006. Discussion note: social and cultural dimensions of air travel demand: Hyper-mobility in the UK? *Journal of Sustainable Tourism*, 14 (2), 209–215.

Sheller, M. 2004. Automotive emotions feeling the car. *Theory, Culture & Society*, 21, 221–242.

Sheller, M. and Urry, J. 2004. *Tourism mobilities: places to play, places in play*. London: Routledge.

Sheller, M. and Urry, J. 2006. The new mobilities paradigm. *Environment and Planning A*, 38, 207–226.

Shoval, N., McKercher, B., Birenbolm, A. and Ng, E. 2015. The application of a sequence alignment method to the creation of typologies of tourist activity in time and space. *Environment and Planning B*, 42 (1), 76–94.

Shove, E. 2003. *Comfort, cleanliness and convenience: the social organization of normal-ity*. Oxford: Berg.

Shove, E. 2010. Beyond the ABC: climate change policy and theories of social change. *Environment and Planning A*, 42, 1273–1285.

Shove, E. 2011. On the difference between chalk and cheese – a response to Whitmarsh et al.'s comments on 'Beyond the ABC: climate change policy and theories of social change'. *Environment and Planning A*, 43, 262–264.

Shove, E., Pantzar, M. and Watson, M. 2012. *The dynamics of social practice: everyday life and how it changes*. London: Sage.

Silverstone, R. (ed.). 1997. *Visions of Suburbia*. London: Routledge.

Simm, B. 2010. *Economic recession and uncertainty; a platform for local transport plan 3 to be innovative*. Report commissioned by the Transport Planning Society.

Simon, H. A. 1955. A behavioral model of rational choice. *Quarterly Journal of Econom-ics*, 69, 99–118.

Sirakaya, E. and Woodside, A. G. 2005. Building and testing theories of decision making by travellers. *Tourism Management*, 26, 815–832.

Slater, D. 1997. *Consumer culture and modernity*. Cambridge: Policy Press.

Slocum, R. 2004. Consumer citizens and the cities for climate protection campaign. *Envi-ronment and Planning A*, 36, 763–782.

Slovic, P. E. 2000. *The perception of risk.* London: Earthscan.

Smith, W., Pitts, R. E. and Litvin, S. W. 2012. Travel and leisure activity participation. *Annals of Tourism Research,* 39 (4), 2207–2210.

Social Exclusion Unit. 2003. *Making the connections – final report on transport and social inclusion.* Social Exclusion Unit – Office of the Deputy Prime Minister, February 2003.

Söderström, O., Paasche, T. and Klauser, F. 2014. Smart cities as corporate storytelling. *City,* 18 (3), 307–320.

Song, Y. and Knaap, G. J. 2004. Measuring urban form: is Portland winning the war on sprawl? *Journal of the American Planning Association,* 70, 210–225.

Soper, K. 2008. Alternative hedonism, cultural theory and the role of aesthetic revisioning. *Cultural Studies,* 22 (5), 567–587.

Soron, D. 2010. Sustainability and the sociology of consumption. *Sustainable Development,* 18, 172–181.

Spaargaren, G. and Mol, A. P. J. 2008. Greening global consumption: redefining politics and authority. *Global Environmental Change,* 18, 350–359.

Spinney, J. 2006. A place of sense: a kinaesthetic ethnography of cyclists on Mt Ventoux. *Environment and Planning D: Society & Space,* 24, 709–732.

Spinney, J. 2008. Cycling between the traffic: mobility, identity and space. *Urban Design Journal,* No. 108 (no page numbers).

Spinney, J. 2009. Cycling the city: movement, meaning and method. *Geography Compass,* 32, 817–835.

Spinney, J. 2010. Performing resistance? Re-reading urban cycling on London's South Bank. *Environment and Planning A,* 42, 2914–2937.

Spinney, J. 2011. A chance to catch a breath: using mobile video ethnography in cycling research. *Mobilities* (special issue), 6, 161–182.

Steg, L. and Gifford, R. 2005. Sustainable transport and quality of life. *Journal of Transport Geography,* 13, 59–69.

Steininger, K. W. and Bachner, G. 2014. Extending car-sharing to serve commuters: an implementation in Austria. *Ecological Economics,* 101, 64–66.

Stern, N. and 22 co-authors. 2006. *Stern review on the economics of climate change.* London: HM Treasury, 30 October 2006.

Stern, P. 2000. New environmental theories: toward a coherent theory of environmentally significant behaviour. *Journal of Social Issues,* 56, 407–424.

Steuteville, R. 2002. The new urbanism: an alternative to modern automobile-orientated planning and development. *New Urban News,* July 2007, 1–5.

Stilgoe, J. R. 1988. *Borderland: origins of the American suburb, 1820–1939.* New Haven, CT: Yale University Press.

Stoll-Kleemann, S., O'Riordan, T. and Jaeger, C. C. 2001. The psychology of denial concerning climate mitigation measures: evidence from Swiss focus groups. *Global Environmental Change,* 11, 107–117.

Stradling, S. G., Meadows, M. L. and Beatty, S. 2000. Helping drivers out of their cars. Integrating transport policy and social psychology for sustainable change. *Transport Policy,* 7, 207–215.

Talen, E. 1999. Sense of community and neighbourhood form: an assessment of the social doctrine of new urbanism. *Urban Studies,* 36, 1361–1379.

Talen, E. 2001. Traditional urbanism meets residential affluence: an analysis of the variability of suburban preference. *Journal of the American Planning Association,* 67, 199–216.

Taylor, D. J. 2000. The sounds of the suburbs: the idea of the suburb in English pop, in *Expanding Suburbia: Reviewing Suburban Narratives*, edited by R. Webster. Oxford: Berghahn. 161–172.

The Telegraph. 2014. *Nick Clegg tells David Cameron to come clean on new garden cities*. [Online]. Available at: www.telegraph.co.uk/news/politics/nick-clegg/10580873/Nick-Clegg-tells-David-Cameron-to-come-clean-on-new-garden-cities.html [accessed: 14th August 2014].

Thaler, R. H. and Sunstein, C. R. 2008. *Nudge: improving decisions about health, wealth and happiness*. New Haven, CT: Yale University Press.

Thompson, F. M. L. (ed.). 1982. *The rise of Suburbia*. Leicester: Leicester University Press.

Tideswell, C. and Faulkner, B. 2002. Multi-destination, tourist travel; some preliminary findings on international visitors' explorations of Australia. *Tourism* (Zagreb), 50 (2), 115–130.

Timmermans, H. J. P. and Zhang, J. 2008. Modelling household activity travel behaviour: examples of state of the art modelling approaches and research agendas. *Transportation Research Part B: Methodological*, 43, 187–190.

TNS. 2013. *Holiday attitudes 2013*. [Online]. Available at: www.tnsglobal.co.uk [accessed: 11th January 2017].

Towner, J. 1996. *An historical geography of recreation and tourism in the western world*. Chichester: John Wiley.

Travel Supermarket. 2013. *Travel Trends Tracker (February) issue 2*. [Online]. Available at: www.travelsupermarket.com [accessed: 14th July 2014].

United Nations Conference on Environment and Development (UNCED). 1992. *Agenda 21 – action plan for the next century*. Endorsed at UNCED. New York: UNCED.

United Nations World Tourism Organization (UNWTO). 2015. *Tourism highlights* (Madrid WTO). [Online]. Available at: www2.unwto.org/publication/unwto-annual-report-2015 [accessed: 6th April 2017].

Universal Productions. 2001. *The fast and the furious*. Universal City, CA: Universal Productions.

Upham, P. 2003. *Towards sustainable aviation*. Edited book. London: Earthscan Publications Ltd.

Urry, J. 1990. *The tourist gaze: leisure and travel in contemporary societies*. 1st edition. London: Routledge.

Urry, J. 2000. *The sociology beyond societies mobilities for the twenty-first century*. London: Routledge.

Urry, J. 2007. *Mobilities*. London: Polity Press.

Urry, J. 2010. Sociology and climate change. *The Sociological Review*, 57 (2), 84–100.

Urry, J. 2011. Does mobility have a future?, in *Mobilities: New Perspectives on Transport and Society*, edited by M. Grieco and J. Urry. Aldershot: Ashgate. 3–19.

Vannini, P. (ed.). 2009. *The cultures of alternative mobilities: routes less travelled*. Aldershot: Ashgate.

Vanolo, A. 2014. Smartmentality: the smart city as disciplinary. *Urban Studies*, 51 (5), 883–898.

Verbeek, D. and Mommaas, H. 2008. Transitions to sustainable tourism mobility: the social practices approach. *Journal of Sustainable Tourism*, 16, 629–644.

Verplanken, B. and Aarts, H. 1999. Habit, attitude, and planned behavior: is habit an empty construct or an interesting case of goal-directed automaticity? *European Review of Social Psychology*, 107, 101–134.

Verplanken, B., Aarts, H. and Van Knippenberg, A. 1997. Habit, information acquisition, and the process of making travel mode choices. *European Journal of Social Psychology*, 27, 539–560.

Viitanen, J. and Kingston, R. 2014. Smart cities and green growth: outsourcing democratic and environmental resilience to the global technology sector. *Environment and Planning A*, 46 (4), 803–819.

Visit Britain. 2016. *The GB day visitor: statistics 2015*. [Online]. Available at: www.visitbritain.org/sites/default/files/vb-corporate/Documents-Library/documents/England-documents/gbdvs_annual_report_2015_13.06.16.pdf [accessed: 6th April 2017].

Wallace Arnold. 1970. *Red Carpet Continental Holidays*. British Library YD.2008 b.68.

Wallace Arnold and Wallace Arnold Air. 1972. British Library YD 2008b.112.

Walt Disney Pictures. 1988. *Who framed Roger Rabbit?* Hollywood, CA: Walt Disney Pictures.

Walton, J. K. 1983. *The English seaside resort: a social history 1750–1914*. Leicester: Leicester University Press.

Walton, J. K. 2011. The origins of the modern package tour? British motor-coach tours in Europe, 1930–70. *The Journal of Transport History*, 32, 145–163.

Wang, D., Xiang, Z. and Fasenmaier, D. R. 2016. Smartphone use in everyday life and travel. *Journal of Travel Research*, 55 (1), 52–63.

Warde, A. 2014. After taste: culture, consumption and theories of practice. *Journal of Consumer Culture*, 14, 279–303.

Warner Brothers. 2015. *Two and a half men*. Burbank, CA: Warner Brothers.

Watt, P. and Smets, P. (eds.). 2014. *Mobilities and neighbourhood belonging in cities and suburbs*. Basingstoke: Palgrave Macmillan.

Weilenmann, A., Normark, D. and Laurier, E. 2013. Managing walking together: the challenge of revolving doors. *Space and Culture*, 17, 122–136.

Wensveen, J. G. and Leick, R. 2009. The Long haul low-cost carrier: a unique business model. *Journal of Air Transport Management*, 15, 127–133.

Whatmore, S. J. 2009. Mapping knowledge controversies: science, democracy and the redistribution of expertise. *Progress in Human Geography*, 33, 587–598.

Whitehead, M., Jones, R. and Pykett, J. 2011. Governing irrationality, or a more than rational government? Reflections on the rescientisation of decision making in British public policy. *Environment and Planning A*, 43 (12), 2819–2837.

Whitmarsh, L., O'Neill, S. and Lorenzoni, I. 2011. Climate change or social change? Debate within, amongst, and beyond disciplines. *Environment and Planning A*, 43, 258–261.

Wiese, A. 2005. *Places of their own: African American suburbanization in the twentieth century*. Chicago, IL: University of Chicago Press.

Williams, S. 1998. *Tourism geography*. London: Routledge.

Wilson, C. and Chatterton, T. 2011. Multiple models to inform climate change policy: a pragmatic response to the 'beyond the ABC' debate. *Environment and Planning A*, 43, 2781–2787.

World Commission on Environment and Development (WCED). 1987. *Our common future*. Report of the World Commission on Environment and Development. Oxford: Oxford University Press.

World Travel and Tourism Council. 2016. *The economic impact of travel and tourism*. [Online]. London: WTTC. Available at: www.wttc.org [accessed: 6th April 2017].

Wright, S. 2002. Sun, sea, sand and self-expression, mass tourism as an individual experience, in *The Making of Modern Tourism*, edited by H. Berghuff, B. Kortie, R. Schneider and C. Harvie. London: Palgrave. 181–202.

Wu, M.-Y. and Pearce, P. 2016. Tourism blogging motivations: why do Chinese tourists create little "lonely planets". *Journal of Travel Research*, 55 (4), 537–549.

Wylie, J. 2009. Landscape, absence and the geographies of love. *Transactions of the Institute of British Geographers*, 34, 275–289.

Wynne, B. 2002. Risk and environment as legitimatory discourses of technology: reflexivity inside out? *Current Sociology*, 50, 459–477.

Yigitcanlar, T. and Lee, S. H. 2014. Korean ubiquitous-eco-city: a smart sustainable urban form or a branding hoax? *Technology Forecasting and Social Change*, 89, 100–114.

Yoo, K.-H. and Gretzel, V. 2011. Influence of personality on travel-related consumer-generated media creation. *Computers in Human Behaviour*, 27, 609–621.

Youtube. 2008a. *1950's TV Ad*. [Online]. Available at: www.youtube.com/watch?v=m7t9YlMxWoE [accessed: 14th August 2014].

Youtube. 2008b. *Harry Enfield: loadsamoney*. [Online]. Available at: www.youtube.com/watch?v=ON-7v4qnHP8 [accessed: 21st June 2017].

Zillinger, M. 2007. Tourist routes: a time-geographical approach on German Car-tourists in Sweden. *Tourism Geographies*, 9 (1), 64–83.

Index

Drawings, graphs and pictures are given in *italics*.

Milton Keynes UK
Ingram Content Group UK Ltd.
UKHW040056071024
449327UK00019B/595

9 780367 362324